AF531456

SOLID STATE CHEMISTRY

SOLID STATE CHEMISTRY

By

M. G. Arora

ANMOL PUBLICATIONS PVT LTD
NEW DELHI-110 002 (INDIA)

ANMOL PUBLICATIONS PVT LTD
4374/4B, Ansari Road, Daryaganj
New Delhi-110 002

Solid State Chemistry

First Edition 1997

ISBN 81-7488-464-5
Reprint, 2001

PRINTED IN INDIA

Published by J. L. Kumar for Anmol Publications Pvt Ltd, New Delhi and Printed at Mehra Offset Press, Delhi.

Contents

Preface

In recent years, few books have been published on solid state chemistry by Indian publishers. The main drawback of all those books is that no book covers complete subject matter. The present book has been written to cover the complete syllabus in such a manner that students don't get confused while going through this book. Each topic included in this book is self-sufficient in itself and has been confined in the light of modern development in a simple and elegant style. Throughout this book, it is assumed that the student understands the fundamental concepts in physical chemistry. Each topic covered in this book can do full justification for most of the students.

I am thankful to my several immediate colleagues and friends for making useful suggestions during the preparation of manuscript of this book.

The author expresses his sincerest thanks to Kiran Pal whose continuous inspiration has initiated the author to bring out this book.

Suggestions for further improvements towards this book shall be gratefully acknowledged and incorporated in the next edition.

—Author

1

Introduction

1.1. Solids

Solids are the substances which maintain their definite shapes against mild distorting forces. Examples of solids are NaCl, KCl, etc. A true solid possesses the following characteristics:

(*i*) A sharp melting point.

(*ii*) A characteristic heat of fusion.

(*iii*) General incompressibility.

(*iv*) A definite three-dimensional arrangement.

1.2. Kinetic Theory and Properties of Solids

The properties of a typical solid are:

1. Definite volume and shape. A solid has a definite volume and shape, neither of which changes appreciably as temperature changes. Gases and liquids have random motion of molecules but the molecular motion in liquids is much less than it is in a gas. On cooling a liquid, its molecules lose energy, their molecular motion decreases and the strong attractive forces draw the molecules closer and fix them in a definite solid structure.

2. Incompressibility. The particles in a solid are practically in contact with each other and so leave very little free space between them. The solids are, therefore, practically incompressible.

3. Vibration of particles and slow diffusion. The atoms, ions or molecules in a solid do not have free motions but they **vibrate or oscillate about** their mean fixed positions and, therefore, cannot detach themselves from the bulk of the solid. The intermolecular forces are so strong that when two solids are placed in contact with each other, they practically do not diffuse into each other.

4. Low vapour pressure. Most solids have low vapour pressures because the molecules in a solid have relatively low kinetic energy. Some solids like iodine, ammonium chloride and camphor have high vapour pressures and change directly into the vapour state on heating.

5. Melting point. Each solid has a characteristic temperature called **melting point** at which it changes into a liquid on heating. When a solid is heated, the particles vibrate rapidly. Further increase in temperature increases their kinetic energies to such an extent that the intermolecular forces can no longer hold the vibrating particles together and thus allow them to slip out of their rigid positions in the solid lattice. The particles are now free to move relatively and constitute the liquid state. A solid melts at constant temperature known as its melting point and the heat absorbed per mole of the solid is called molar heat of fusion of the solid.

Table 1.1. Melting Points of a few Substances

Solid	*Melting point*	*Solid*	*Melting point*
Oxygen	55 K	Sodium	371 K
Nitrogen	63 K	Sodium chloride	1077 K
Ethyl alcohol	159 K	Magnesium chloride	1200 K
Carbon tetrachloride	249 K	Diamond	3773 K

Melting point of a solid gives an idea of the strength of binding forces *(Intermolecular attractive forces)* between its atoms, ions or molecules. For example, melting point of sulphur is 392 K and that of diamond is 3773 K. This shows that the binding forces are much stronger in diamond than those in sulphur.

1.3. Classification of Solids

Many solids are hard and others are soft. Diamond is a hard solid while graphite is soft. Some solids are good conductors and others are poor or bad conductors. They also differ in melting points. Metals usually have high melting points whereas phosphorus has low melting point. **It is clear that solids not only differ from liquids and gases but also differ from one-another in many of their properties.** The reason for this is the different types of arrangements possible for atoms, ions and molecules and the different nature and strength of the forces acting between them.

Table 1.2. Distinction Between Solids, Liquids and Gases

Property	*Solids*	*Liquids*	*Gases*
1. Nature	Consists of atoms, molecules or ions	Consists of molecules	Consists of molecules *e.g.*, H_2, I_2. Noble gases are monoatomic
2. Shape	Definite shape	No definite shape	No definite shape
3. Volume	Definite volume	Definite volume	No definite vol.
4. Positions of atoms, molecules or ions	Positions of atoms, molecules or ions are fixed	Positions of molecules are not fixed	Positions of molecules are not fixed
5. Molecular Distances	Distances between molecules, atoms or ions are small	Inter-molecular distances are of intermediate sizes	Large intermolecular distances
6. Compres-sibility	Not easily compressible	Can be compressed slightly	Can be compressed
7. Density	High	Intermediate	Low
8. Brownian motion	No Brownian motion, the constituent particles vibrate about mean positions	Show Brownian motion to a lesser extent	Full Brownian motion
9. Kinetic energies	Low	Intermediate	High
10. MP's and BP's	Usually high. Low in case of molecular solids	Intermediate	Low
11. Vapour pressure	Low	High	High

Solids may be roughly classified into two types: **True solids and pseudo solids.** A true solid has a definite shape and definite volume. A true solid is rigid and is not easily distorted by mild distorting forces. A pseudo solid loses shape on long standing a flows under its own weight and is easily distorted by even mild distortion forces. Glass and pitch are pseudo solids. Pseudo solids are called super cooled liquids.

Table 1.3. Distinction between True Solids and Pseudo-solids

Property	*True solids*	*Pseudo solids*
1. Shape and volume	Have definite shapes and definite volumes. Do not lose shape even on long standing.	No definite shape and volume. On long standing, they lose shape and volume and flow under their own weights.
2. Figidity and resistance to distortion forces	More rigid and are not easily distorted by mild distortion forces *e.g.* all crystalline solids (KCl, NaCl, Fe, Cu, S, sugar etc.)	Less rigid and easily distorted by even mild distortion forces *e.g.*, glass and pitch.
3. Melting point	Sharp melting point	Melt over a wide range of temperature. They usually first soften and then pass to the liquid state.

Solids may also be classified into the following two types:

(i) Amorphous solids

(ii) Crystalline solids.

(a) Amorphous solids. These are such solids which have incompressibility and rigidity but they do not have a geometrical shape. In amorphous solids, though the atoms or molecules are strongly bonded yet there is no geometrical regularity or periodicity in the way in which atoms are arranged in space. Examples of amorphous solids are glass, fused silica, rubber and polymers of high molecular masses.

(b) Crystalline solids. Some solids have incompressibility and rigidity but they possess a definite geometrical shape. These solids are known as *crystalline solids*. A crystalline solid possesses a definite and regular geometry due to definite and orderly arrangement of molecules, atoms or ions in a three dimensional space. A crystalline solid has **same geometry** irrespective of the source from which it is obtained. Some examples of crystalline solids are NaCl, KCl, etc.

1.4. Distinction between Crystalline and Amorphous Solids

(a) Characteristic Geometry. Crystalline solids have a regular arrangement of particles in typical geometrical forms, whereas amorphous solids have a complete random particle arrangement.

(b) Melting Point. A crystalline substance has a very sharp melting point while an amorphous substance does not have. For example, as the temperature of glass is raised gradually, it softens and then starts flowing without underging any abrupt or sharp change from solid to liquid state.

(c) Cooling Curve. Cooling curve for an amorphous substance is smooth, while the curve for a crystalline substance has two

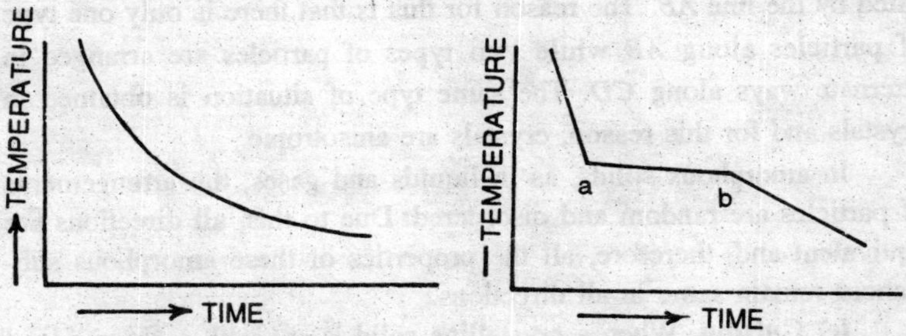

Fig. 1.1. Amorphous substance. Fig. 1.2. Crystalline substances.

breaks (a) and (b) which correspond to the beginning and the end of the process of crystallization. Temperature remains constant during crystallization. The process of crystallization is accompanied by some liberation of energy, which compensates for the loss of heat and causes the temperature to remain constant (Figs. 1.1 and 1,2).

(d) Isotropy and Anisotropy. Amorphous substances are said to be *isotropic*, if their properties such as electrical conductivity, thermal conductivity, mechanical strength and refractive index are same in all directions. Crystalline solids, on the other hand, are *anisotropic, i.e.* their physical properties are different in different directions.

The phenomenon of anisotropy offers a strong evidence for the presence of ordered molecular arrangements in crystals. This can be understood from Fig. 1.3 in which a simple two dimensional arrangement of only two different types of atoms has been depicted. If the

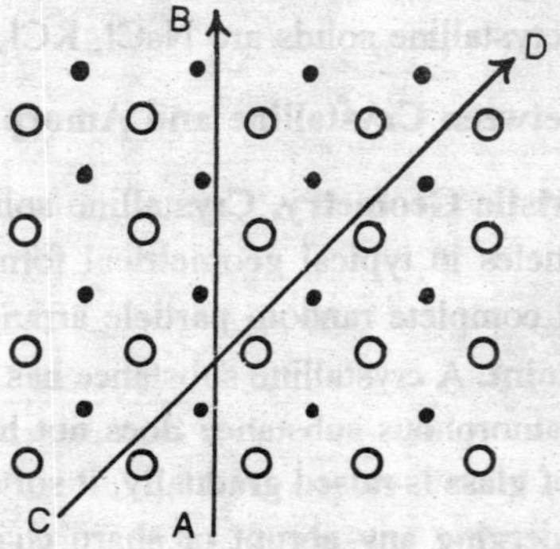

Fig. 1.3. Anisotropic behaviour of crystals.

properties are measured along *AB* and *CD*, the properties measured along *CD* will be different from those measured in the direction indicated by the line *AB*. The reason for this is that there is only one type of particles along *AB* while two types of particles are arranged in alternate ways along *CD*. The same type of situation is obtained in crystals and for this reason, crystals are anisotropic.

In amorphous solids, as in liquids and gases, the arrangements of particles are random and disordered. Due to this, all directions are equivalent and, therefore, all the properties of these amorphous substances remain same in all directions.

(e) Cutting. When a crystalline solid is cut with a sharp edged tool, it yields a clean cleavage. [Fig. 1.4 (a)]. On the other hand, when an amorphous solid is cut with a sharp edged tool, it causes conchoidal fracture [Fig. 1.4 (b)].

Fig. 1.4.. The Cutting of solids *t (a)* A crystalline solid gives a clean cleavage *(b)* An amorphous solid gives an irregular cut.

Table 1.4. Distinction between Crystalline and Amorphous Solids

Property	*Crystalline solids*	*Amorphous solids*
1. Structure	The atoms, ions or molecules are arranged in three dimensional space in a definite and regular manner and therefore have definite and regular geometry.	The atoms, ions or molecules are arranged in three dimensional space in an indefinite and irregular manner and therefore do not have definite geometric shapes.
2. Cutting with a sharp edged tool (*cleavage*)	Clean cleavage, *i.e.*, it breaks into two pieces with plane surfaces.	Unclean cleavage *i.e.*, it breaks into two pieces with irregular surfaces.
3. Compressibility	Rigid and incompressible *e.g.* Sodium chloride, diamond, sugar, sulphur etc.	Usually rigid and cannot be compressed to any appreciable extent *e.g.*, rubbers, certain plastics, glass, fused silica. Graphite is soft because of its unusual structure. These substances are, often micro-crystalline and on atomic scale contain regular repeating units.
4. Melting point	A definite (sharp) melting point.	Melt over a wide range of temperature.
5. Heat of fusion	A definite heat of fusion.	No definite heat of fusion.
6. Anisotropy	Anisotropic *i.e.*, they have unequal physical properties along different axes. This shows that there is long range orderly arrangement of atoms, ions or molecules in crystals.	Isotropic *i.e.*, they have same physical properties in all directions like electrical conductivity, refractive index, thermal expansion etc.

1.5. Crystal

A crystal is a solid composed of atoms, molecules or ions arranged in an orderly repetitive array. In other words, a crystal may also be defined as a homogeneous anisotropic substance having a definite geometrical shape within well defined surfaces known as planes. A crystal also has the following characteristics:

(*i*) A sharp melting point,

(*ii*) A characteristic heat of fusion,

(*iii*) A definite three-dimensional arrangement of constituent particles, and

(*iv*) general incompressibility.

1.6. External Features of Crystals

The external geometrical form of a crystal is obtained as a result of internal regular arrangements of atoms of which it is built up. The regularity is that of a three dimensional pattern in which a certain unit of structure is repeated over and over again in space. The different external features of crystals are briefly described below:

(a) Faces. *Crystals are bounded by a number of surfaces which are generally perfect flat. These surface are known as faces.* Faces are of two types: *like* and *unlike*. There are some crystals which are bounded by faces which are all alike while there are some crystals which are bounded by different faces. For instance, fluorspar is generally obtained in cubes, alum in regular octahedron whereas galena crystal exhibits a combination of the cube and an octahedron (Fig. 1.5).

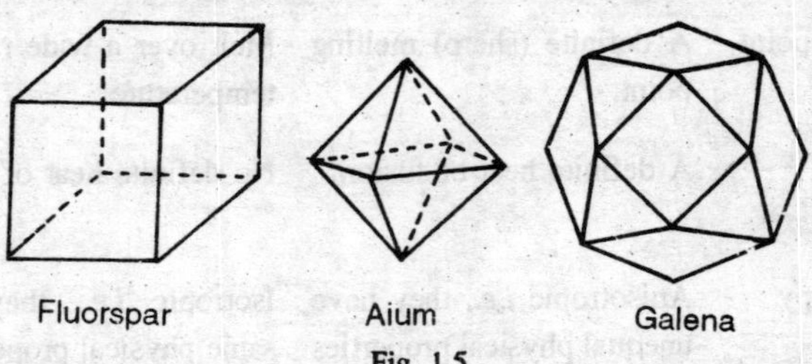

Fig. 1.5

(b) Form. *All the faces corresponding to a crystal are said to constitute a 'form'*. A simple form of a crystal is entirely made up of like faces whereas a crystal which consists of two or more simple forms is known as a *combination.*

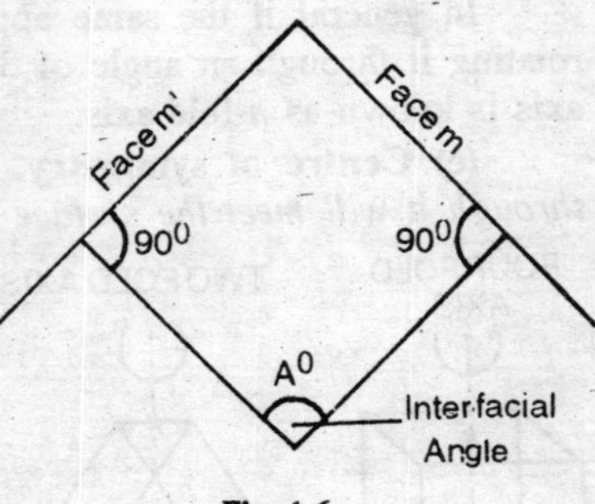

Fig. 1.6

(c) Edges and interfacial angles. An edge is formed by the intersection of two adjacent faces whereas the angle between any two faces of a crystal is known as the interfacial angle. The crystals are bounded by plane faces, straight edges and interfacial angles. The relationship between these elements can be expressed by the formula:

$$f + c = e + 2$$

where f, c and e denote the number of faces, angles and edges respectively.

(d) Zone axis. A line drawn through the centre of a crystal in a direction parallel to the edges of a zone is called the *zone axis.*

1.7. Symmetry of Crystals

In order to understand symmetry, we shall first consider the kinds of *symmetry elements* a molecule may possess and the *symmetry operations* generated by symmetry elements.

Symmetry elements. A symmetry element is a *geometrical entity such as a line (or axis), a plane or a point with respect to which one or more symmetry operations may be carried out.* The various types of elements are as follows:

(a) Plane of symmetry. *A crystal is said to possess as plane of symmetry when an imaginary plane passing through the centre of crystal can divide it into two parts such that one is the exact mirror image of the other.* The standard rotation for a plane of symmetry is indicated by σ.

(b) Axis of symmetry. *It is a line about which the crystal may be rotated so that it represents the same appearance more than once during a complete revolution.* If the equivalent configuration occurs twice, thrice, four and six times, *i.e.*, after rotation of— 180°, 120°, 90° and 60°, the axes of rotation are known as two fold (diad), three-fold (triad), four-fold (tetrad) and six-fold (hexad), axes of symmetry respectively. The axis of symmetry is indicated by C. If it is two fold it is indicated by C_2. Similarly, three-fold and six-fold axes of symmetry are indicated by C_3 and C_6 respectively (Fig. 1.7).

In general if the same appearance of a crystal is repeated on rotating it through an angle of $360°/n$ around an imaginary axis the axis is known as n-fold axis.

(c) Centre of symmetry. *It is a point that any line drawn through it will meet the surface of the crystal at equal distances on*

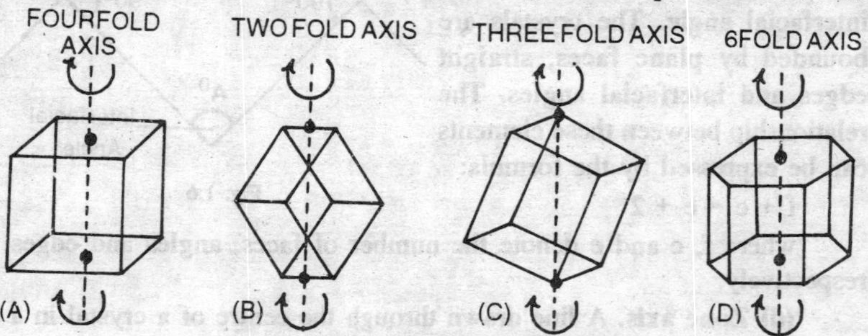

Fig. 1.7 Various axis of symmetry

either side. It is important to mention here that a crystal may possess a number of planes or axes of symmetry but it can only have one centre of symmetry.

1.8. Symmetry Operation

A symmetry operation is a movement of a body, such that, after the movement has been carried out, every point of the body is coincident with an equivalent point (or perhaps the same point) of the body in its original orientation. There are four principal operations for repeating a figure:

(a) Translation operation (b) Rotation operation (c) Reflection operation across a line in two dimensions or plane in three dimensions and (d) Inversion through a point (Fig.1.8).

Symmetry Elements of a Cubic Crystals. A cube has thirteen axes of symmetry (three four-fold, four three-fold, six two-fold), nine planes of symmetry and one centre of symmetry *i.e.*, 23 elements symmetry altogether. This will be under stood from the discussion given below:

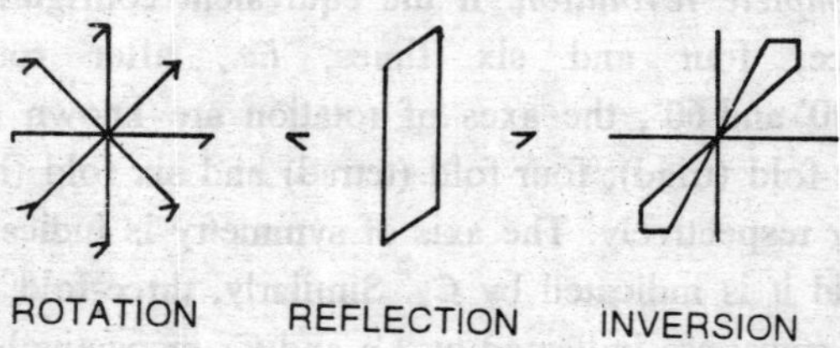

Fig. 1.8

(a) Rectangular planes of symmetry. In Fig. 1.9 (a), one rectangular plane of symmetry has been shown. Besides, this, there are two more such planes not shown in Fig. 1.9 (a). Each of these two planes is at right angle to the plane shown in Fig. 1.9. Hence, the total number of rectangular planes will be **three.**

(b) Diagonal plane of symmetry. In Fig. 1.9 (b), one plane passing diagonally through the cube has been shown. Similar to this, there are five more such planes passing diagonally through the cube. Hence, the total number of planes passing diagonally through the cube will be six.

(c) Centre of symmetry. Only one centre of symmetry is possible which is lying at the centre of cube. It is shown in Fig.1.9 (c).

(d) Axes of two-fold symmetry. In Fig. 1.9 (d), one axis of two- fold symmetry emerging from opposite edges has been shown. Evidently, there will be a total of six such axes of two-fold symmetry.

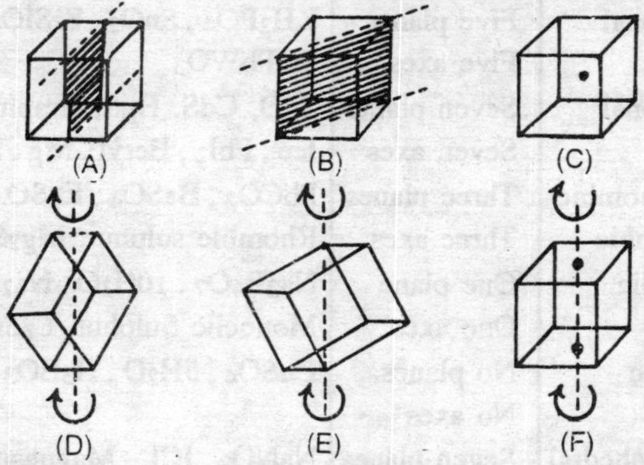

Fig. 1.9. Various elements of symmetry in a cube.

(e) Axes of three-fold symmetry. In Fig. 1.9 (a), one axis of three-fold symmetry passing through opposite corners has been shown. Evidently, there will be a total of **four** such axes.

(f) Axes of four-fold symmetry. In Fig. 1.9 (f), one axis of four-fold symmetry has been shown. Evidently, there will be a total of three such four-fold axes at right angles to one another.

From the above discussion the number of symmetry elements of various types in a cube would be 23, *i.e.,*

Centre of symmetry = 1

Planes of symmetry = 3 + 6 = 9

Axes of symmetry = 3 + 4 + 6 = 13.

∴ Total number of elements will be 1+ 9 +13 or 23.

1.9. Point Groups

From geometrical considerations, it may be assumed that 32 different combinations of elements of symmetry of a crystal are possible theoretically. These are known as 32 **point groups** or 32 **systems.** But some of these could be grouped together. Therefore, these 32 systems could be grouped together in seven different categories which are known as the **seven basic crystal** systems. These seven systems are *cubic* or *regular, tetragonal, hexagonal, orthorhombic* or *rhombic, monoclinic, triclinic* and *rhombohedral* or *trigonal.* All these systems with the maximum number of *planes* and *axes* symmetry elements have been given in the following Table.

System	*Symmetry*	*Examples*
1. Cubic or Regular	Nine planes Thirteen axes	$NaCl$, KCl, CaF_2, Cu_2O, ZnS Pb, Ag, Au, Hg, alums, diamond
2. Tetragonal	Five planes Five axes	KH_2PO_4, SnO_2, $ZrSiO_4$, TiO_2 Sn,$PbWO_3$
3. Hexagonal	Seven planes Seven axes	ZnO, CdS, HgS, Graphite, Ice, PbI_2, Beryl, Mg, Zn, Cd
4. Orthorhombic or rhombic	Three planes Three axes	$PbCO_2$, $BaSO_4$, $K_2SO_4KNO_3$ Rhombic sulphur, Mg_2SiO_4
5. Monoclic	One plane One axes	$Na_2B_4O_7 . 10H_2O . Na_2SO_4 . 10H_2O$ Monoclic Sulphur, $C_aSO_4, 2H_2O$
6. Triclinic	No planes No axes	$CuSO_4$, $5H_2O$, H_3BO_3
7. Reombohedral (or Trigons)	Seven planes Seven axes	$NaNO_2$, ICl, Magnesite, As, Sb, Bi, Quartz

1.10 The Crystal Lattice and Unit Cell

A crystal is a solid particle with well defined planar surfaces known as faces. The shape of a crystal reflects its orderly internal structure. The crystals of the same substance may have different sizes because different faces grow at different rates but the angles between the faces known as interfacial angles remain the same.

Space lattice. A space lattice is an arrangement in space of isolated points in regular pattern that show the positions of atoms, mole~~les or ions in a crystal. The space lattice may also be defined as a regular three dimensional arrangement of equivalent-points in space that shows the geometry of a crystal. In non-crystalline solids,

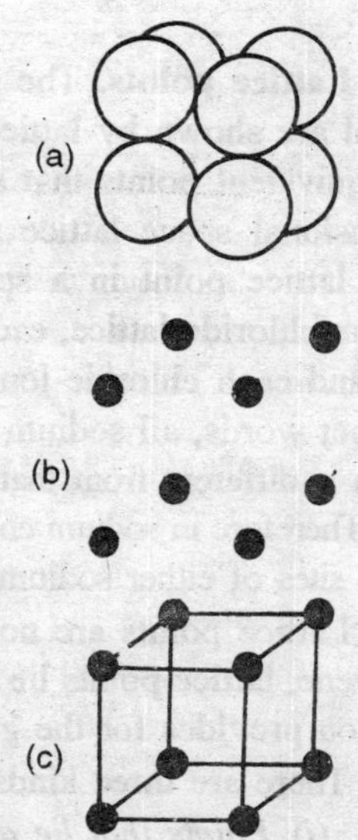

Fig. 1.10. The unit cell of a crystal

the space lattice is distorted and in truly amorphous solids, there is no order.

The unit cell and crystal parameters. The unit cell is the smallest portion of a space lattice which on moving equal distances in various directions reproduces the whole crystal structure. The unit cell is formed by connecting a lattice point with the three nearest lattice points which are not co-plannar with it. This means that unit cell has three lengths and three angles. The crystals of the same substance may have different sizes. Some are small and others are large. For a crystal of the same substance, the three angles remain the same and the lengths may vary. The relative positions of lattice points, lengths and angles are best described by a set of three axes called *crystallographic axes*. In Fig. 1.11. *x*, *y* and *z* are three axes, the intercepts on these are designated by *a*, *b* and *c* respectively. The angle between *y* and *z* axis is α ; heat between *x* and *z* axes is β ; and that between *x* and *y* axes is γ.

Conditions for choosing the unit cell:

(*i*) The unit cell should have the same symmetry as the crystal itself.

(*ii*) Where more than one simple arrangements are possible, the one with the smaller number of atoms or particles is chosen.

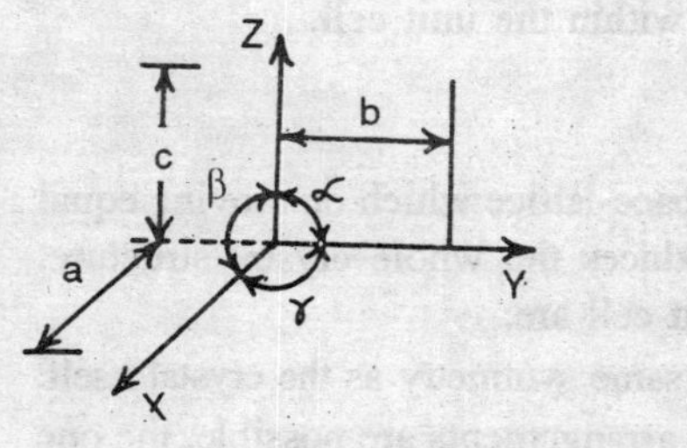

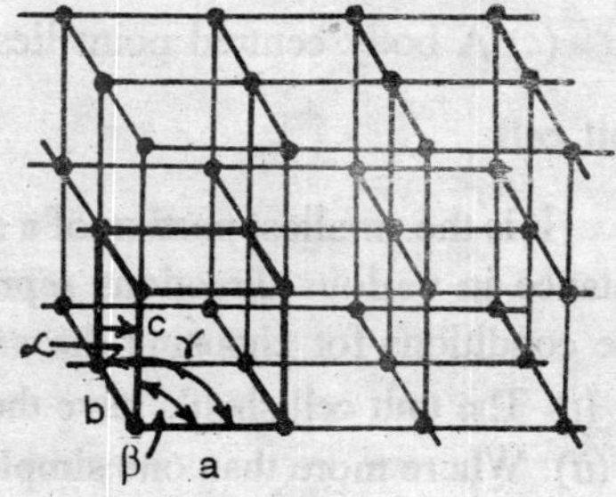

Fig. 1.11. Three dimensional space lattice.

Lattice points. The positions of atoms, molecules or ions in a crystal are shown by lattice points in space lattice. The lattice points are equivalent points that are arranged in a regular pattern in a three dimensional space lattice which shows the geometry of the crystal. Each lattice point in a space lattice has the same environment. In sodium chloride lattice, each sodium ion is surrounded by six chloride ions and each chloride ion in turn is surrounded by six sodium ions. In other words, all sodium ions have the same indentical environment which is different from that of the identical environment of the chloride ions. Therefore in sodium chloride lattice, the lattice points, may be chosen at the sites of either sodium ions or chloride ions but not of both.

Lattice points are not always occupied by atoms. In crystalline benezene, lattice points lie in the centres of benzene rings because this location provides for the greatest symmetry about the lattice points.

There are three kinds of lattice points in a unit cell :

(*i*) *Points that lie at the corners of the unit cell*

(*ii*) *Points that lie at the faces of the unit cell*

(*iii*) *Points that lie within the unit cell.*

(*a*) In a crystal made up of cubic unit cells, each point in the lattice is at the corner of eight adjoining cubes. Therefore, only 1/8 of this point lie within any unit cell (Fig. 1.12).

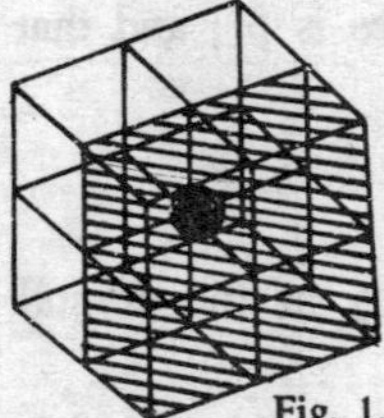

Fig. 1.12. (a)

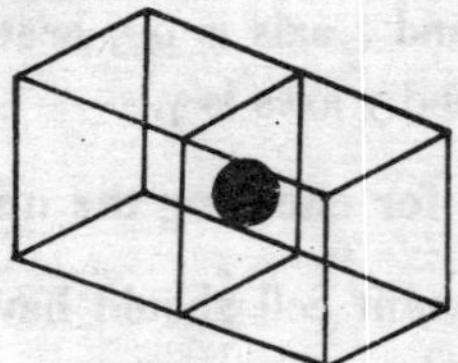

Fig. 1.12. (b)

(*b*) A point in a lattice made up of face-centred unit cells belongs to two cells in the lattice. The atom is in the face common to two cubes.

(*c*) A body centred point lies within the unit cell.

Unit cell

It is the smallest portion of a space-lattice which on moving equal distance in various directions reproduces the whole crystal structure. The conditions for choosing the unit cell are:

(*i*) The unit cell should have the same symmetry as the crystal itself.

(*ii*) Where more than one simple arrangements are possible, the one with the smaller number of atoms or particles is chosen.

In Fig. 1.13(*a*) a two-dimensional geometric pattern is shown. In this there are three possible unit cell marked A, B and C. It is possible to generate space lattice by placing any one of these unit cells side by side in space. But a single unit cell is to be chosen which may be found out by considering the following facts:

(*i*) The symmetry of each of these three unit cell is considered, and

(*ii*) The volume per unit cell is also considered.

Unit cell A is having 2-fold symmetry and is having two dots and one cross per unit cell. Unit cell B is having 4-fold symmetry and is having four dots and two crosses. However, it is larger than A. Unit cell C is also having maximum symmetry of 4-fold axis of rotation and is having two dots and one cross per unit cell. On the basis of minimum volume, cell B could be rejected. Out of cell A and C (both are having the same volume), unit cell A has been rejected (in favour of B) because it is having low symmetry.

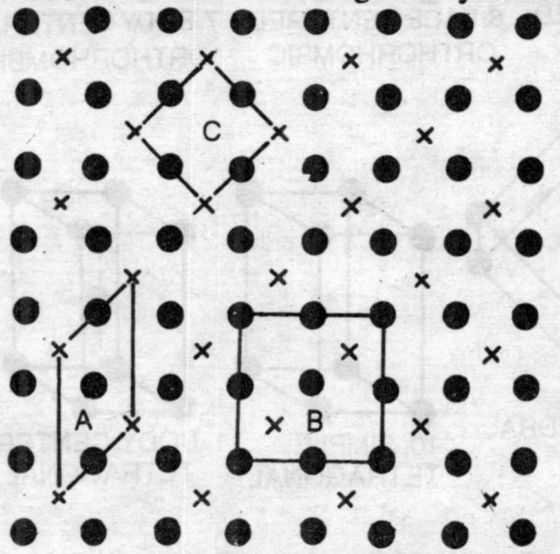

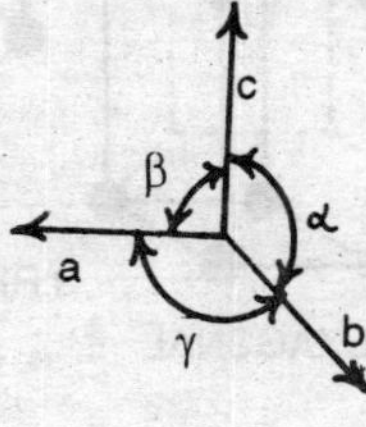

(*a*) Lattice and a unit cell Fig. 1.13 (*b*) Description of a unit cell

In order to describe a unit cell, one has to know:

(*i*) the distance *a, b* and *c* which give the length of the edges of the unit cell and

(*ii*) the angles α, β and γ which give the angles between the three imaginary axes *OX, OY* and *OZ* Fig. 1.13(*b*).

Bravais Lattice. From geometrical considerations it was shown by A. Bravais in 1848 that all possible three-dimensional space lattices are of fourteen distinct types. These are known as *Bravais lattices* and are derived from seven crystal systems. All crystalline solids can be represented by one of these lattice structures. The unit cells for

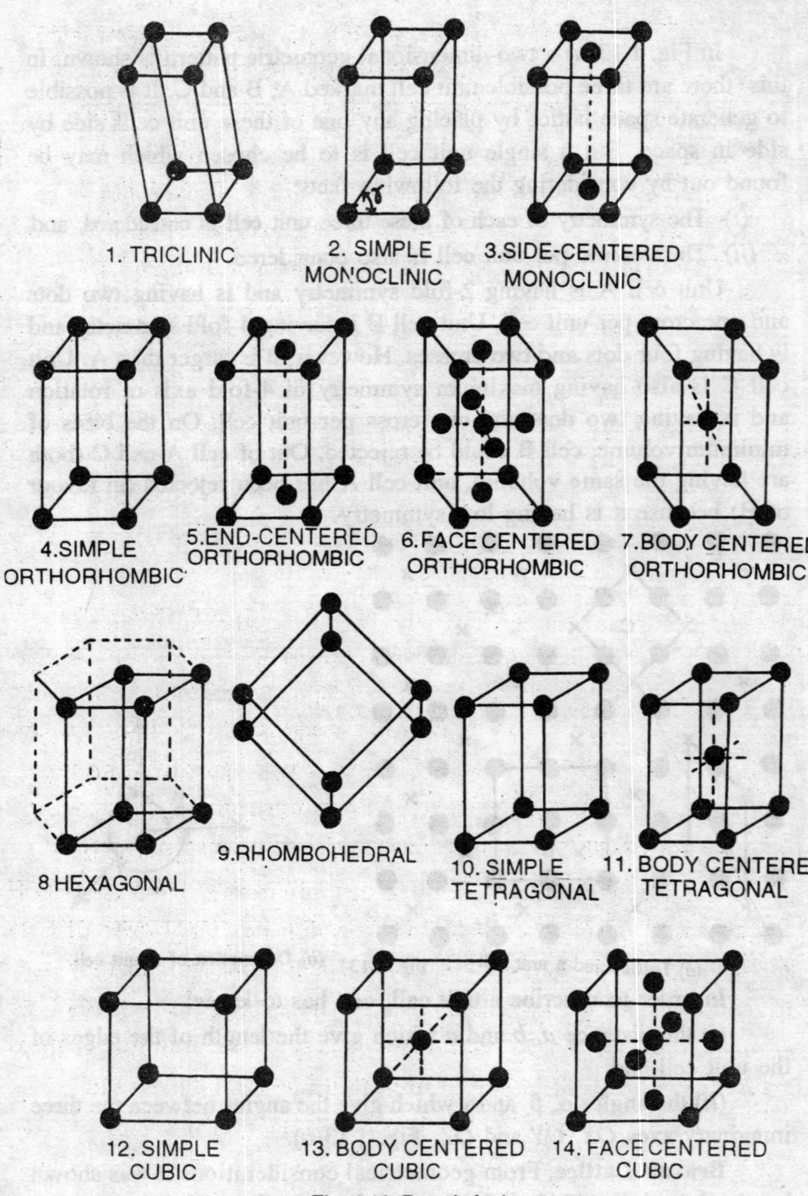

Fig. 1.14. Bravais lattice.

these fourteen Bravais lattice have been shown in Fig. 1.14. In this figure, the parameters of unit cell, *i.e.*, the intercepts *a, b* and *c* and interfacial angles α, β and γ have been shown.

Three types of unit cells for cubic crystals are as follow:

(a) Simple cubic lattice. When one unit is situated only at each corner of the cubic cell, it is known as simple cubic lattice. In other words, it contains eight points which are situated at eight corners of a cube. There is no lattice point inside the cube, [(Fig. 1.14-(12)].

(b) The body centred cubic bcc lattice. When one unit is situated at each corner and one at the centre of a cube, it is known as body centred cubic lattice. In all *bcc* lattice consists of nine points [(Fig. 1.14-(13)].

(c) Free entered cubic bcc lattice. When one unit is situated at each corner and one at the centre of each of the faces, it is known as face centred cubic lattice. In all *fcc* lattice will consist of fourteen points [(Fig. 1.14-(14))].

We have already described that a maximum of 32 elements of symmetry (point groups) would be possible. On combining these with 14 Bravias lattices, 230 different arrangements known as **space groups** are possible.

Classification of Crystals on the Basis of Their Shapes. Theoretically, 230 crystal forms are possible. Practically all have been observed. On the basis of their symmetry, these 230 crystal forms have been divided into 32 classes and these in turn have been further classified into seven crystal systems. The shapes of these are shown in Fig. 1.14. and their characteristics have been given in Table 1.4.

Table 1.5

Crystal system (I)	*Type of lattices (II)*	*Axial distances (III)*	*Axial angles (IV)*	*Examples (V)*
1. Cubic	Primitive, Face-centered, Body-centred = 3	$a = b = c$	$\alpha = \beta = \gamma = 90^\circ$	Copper, KCl and zinc blende
2. Tetragonal	Primitive Body-centred = 2	$a = b \neq c$	$\alpha = \beta = \gamma = 90^\circ$	White tin, SnO_2

3. Ortho-nhombic	Primitive, Face-centred Body-centred End-centred = 4	$a \neq b \neq c$	$\alpha = \beta = \gamma = 90°$	Rhombic sulphur
4. Rhom-bohedral	Primitive = 1	$a = b = c$	$\alpha = \beta = \gamma \neq 90°$	Calcite
5. Mono-clinic	Primitive End-centred = 2	$a \neq b \neq c$	$\alpha = \gamma = 90°$, $\beta \neq 90°$	Monoclinic sulphur
6. Triclinic	Primitive = 1	$a \neq b \neq c$	$\alpha \neq \beta \neq \gamma \neq 90°$	Potassium dichromate
7. Hexagonal	Primitive = 1	$a = b \neq c$	$\alpha = \beta = 90°$, $\gamma = 120°$	Graphite

In Table 1.5 :

(*i*) I column is giving the name of the crystal system.

(*ii*) II column is giving the number and types of space lattices present in the given crystal system.

(*iii*) III column is giving axial distances.

(*iv*) IV column is giving axial columns.

(*v*) V column is giving examples.

1.11. Law of Rational Indices

This law expresses that the intercepts of any face of a crystal along the crystallographic axes are either equal to the unit intercepts (a, b, c) or some simple whole number multiples of them *e.g.*, na, *n*′b, n″c, where n, n′ and n″ are simple whole numbers.

Let OX, OY and OZ represent the three crystallographic axes and ABC and be a unit plane (Fig. 1.15). The unit intercepts will thus be a, b and c on the three axes respectively. According to the above mentioned law, the intercepts of any face such as ABC on the same three axes would be simple whole number multiples of a, b and c respectively. As can be seen from the figure, the simple multiple in this case are 2, 2 and 3.

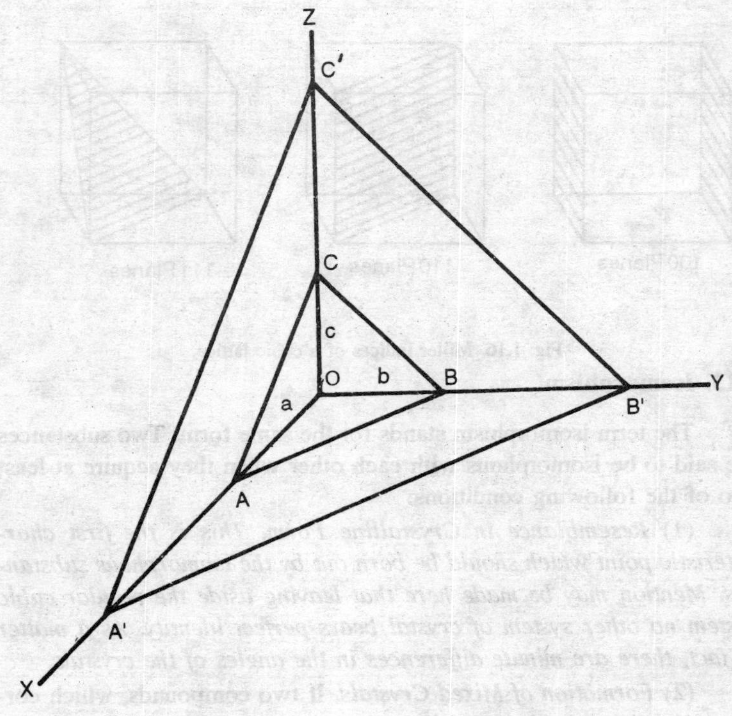

Fig. 1.15. Law of rational indices and crystallographic axes

1.12. Miller Indices

The Miller indices of a given crystal face are the ratio of the reciprocals of the intercepts of this very face on the chosen axes.

For the face A B C the reciprocal of the intercepts on the three axes are $\frac{1}{2}, \frac{1}{2}$ and $\frac{1}{3}$. Obviously the reciprocals are in the ratio of 3 : 3 : 2. The numbers 3 : 3 : 2 are the Miller indices of the face ABC which is rendered as a (332) face. Now for the face ABC (*i.e.*, the unit plane), the intercepts a, b and c are all unity; evidently the reciprocals are in the ratio 1 : 1 : 1. Hence the Miller indices are (111). When a face cuts only two of the axes and is parallel to third, the intercept on that axis would be infinity and the corresponding index is zero. Thus, a plane passing through A and B is parallel to the axis OZ would have the indices (110). Following are the examples of the (100), (110) and (111) planes of a cubic lattice which describe the lattice structure.

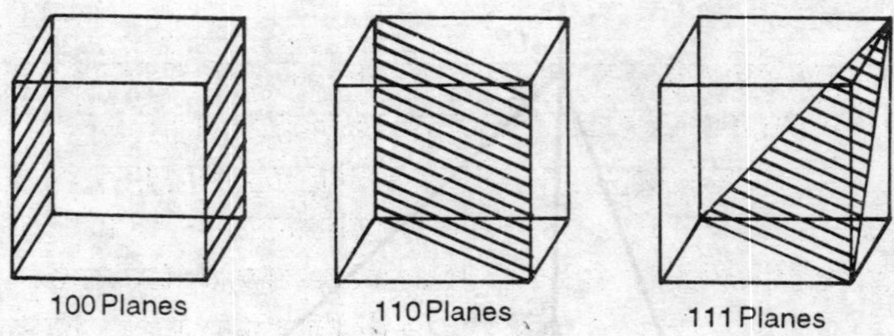

Fig. 1.16. Miller indices of a cubic lattice.

1.13. Isomorphism

The term isomorphism stands for the same form. Two substances are said to be isomorphous with each other when they acquire at least two of the following conditions:

(1) *Resemblance in Crystalline Form. This is the first characteristic point which should be born out by the isomorphous substances. Mention may be made here that leaving aside the regular cubic system no other system of crystal bears perfect identity. As a matter of fact, there are minute differences in the angles of the crystals.*

(2) *Formation of Mixed Crystals.* If two compounds, which correspond to isomorphism are allowed to crystallise from the same solution, the resulting crystals contain both the substances. Another remarkable feature is that the composition for the crystals with respect to the two salts changes continuously. Thus from a solution containing $Cr_2(SO_4)_3 Al_2(SO_4)_3 24H_2O$ and $K_2SO_4Al_2(SO_4)_3 24H_2O$, we get crystals of varying colour from deep violet to pale pink, according as larger or a smaller share of chrome alum is present.

(3) *Formation of Isomorphous Growths.* If a crystal of one of the isomorphous substances is placed in a saturated solution of that substance with which it is isomorphous, the crystal starts growing slowly. Consider, for example, barium sulphate and potassium permanganate, the latter salt forms overgrowth on the former.

However, we hardly find the fulfilment of condition (2) and (3), as there are numerous examples of isomorphous pair of salts where one, or even both of these can fail.

In 1819 Mitscherlich put forward his prediction that isomorphous compounds generally possess similar chemical formulae. Thus, alums which are isomorphous with each other have similar chemical structure and align with them, and have the same number of water molecules of crystallization. (*e.g.,* $Cr_2(SO_4)_3\, Al_2(SO_4)_3 . 24H_2O$ and $K_2SO_4 . Al_2(SO_4)_3 . 24H_2O$). However, it may be noted that the chemical compounds which have the similar formulae do not essentially have the same crystalline form. For example, KNO_3 and $NaNO_3$ are not at all isomorphous. In several cases the same substance may exhibit two different crystalline forms *e.g.,* calcium carbonate as *calcite* and *aragonite.* As calcite it is isomorphous with sodium nitrate while as aragonite it is isomorphous with potassium nitrate.

The chlorate of sodium is not isomorphous with either its nitrate or corbonate, and possesses exclusively different type of structure, since chlorate is pyramidal while others are planner. The essentiality for ions to have similar stereochemical form imposes another condition for isomorphism, apart from the three already mentioned.

Polymorphism

The term poly always stands for more than one and in the study of the crystal structure the term polymorphism is used when the same substance can exhibit more than one crystalline form, for example, calcium carbonate which crystallises, out as *calcite* and *aragonite.* The elements, carbon, phosphorus and sulphur also correspond to this phenomenon and in the case of elements this property is assigned the term allotropy while elements exhibiting this property are known as allotropic elements and their various forms are known as allotropic forms.

1.14. Classification of Crystals on the Basis of Bond Type

On the basis of units, types of binding forces and their properties, the crystals can be divided into the following four categories :

(a) Ionic crystals. The forces which hold the ions together within the ionic crystal are the strong electrostatic coulombic forces. The type of lattice with which the ionic compound crystallises depends on two factors:

(*i*) The size of the ion.

(*ii*) The necessity for the preservation of electrical neutrality. Thus, the lattices in ionic crystals consist of alternate positive and negative ions in equivalent amounts (Fig. 1.17)

The various characteristics of ionic crystals are as follows:

(*i*) These crystals are usually brittle and hard because the electrostatic attractions between oppositely charged ions is fairly strong. They have high melting points. For example, Na_2SO_4, an ionic crystal has the m. p. 1157 K.

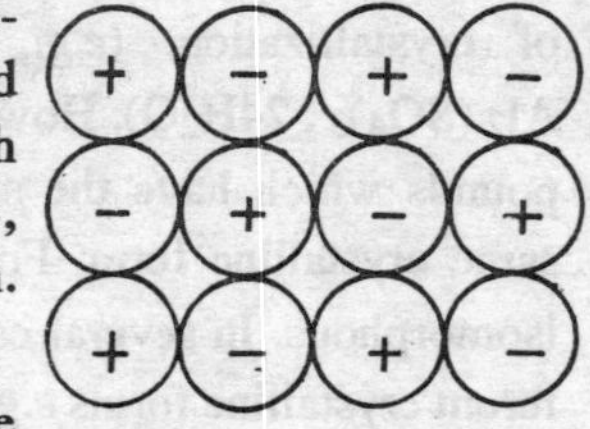

Fig. 1.17

(*ii*) The ionic crystals are often quite soluble in ionising solvents such as water and also in the other polar solvents. They are insoluble or very slightly soluble in non-polar solvents such as chloroform, carbon tetrachloride and benzene.

(*iii*) The ionic crystals usually crystallise in relatively closed structure.

(*iv*) The electrical conductivity of ionic crystals is much lower than that of a metal and the conductivity increases with increasing temperatures. Ionic solids are poor conductors of electricity in the solid state because the ions are entrapped in fixed places in the crystal lattice and will not move under the influence of electric field. However they become good conductors in the molten state because there occurs the destruction of well ordered arrangement of ions in the molten state and therefore the ions will move under the influence of an electric field.

(*v*) The heats of vaporisation of ionic crystals are very high while their vapour pressures at low temperatures are very low.

(*vi*) Ionic crystals are hard and brittle. The reason for this is movement of layers of ions of the same charge near each other and this causes strong repulsion, thereby leading to the breakdown of a crystal (Fig. 1.18).

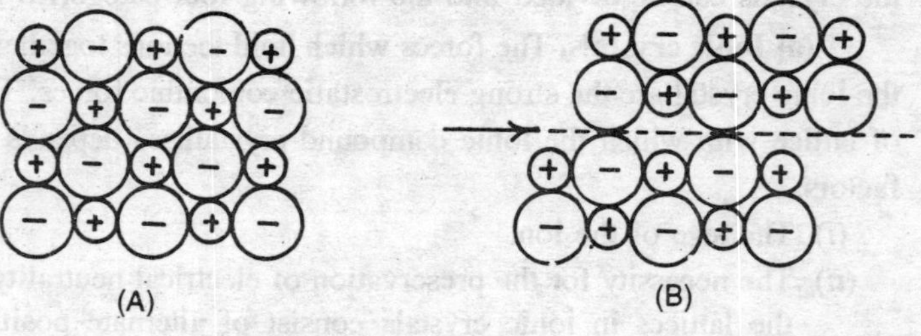

Fig. 1.18. Breaking of a crystal.

Examples of ionic crystals are Nacl, NaF, AgCl and LiF. Ionic solids are formed when electrons from highly electro-positive atoms are completely transferred to electronegative atoms. *An ionic solid contains a large number of positive and negative ions which are arranged in space as to produce maximum stabilization.* Since ionic bond is non-directional, the structure of an ionic solid is determined by purely geometrical considerations. The simple ionic solids have close-packed structures but co-ordination number is rarely more than eight. Majority of ionic solids are cubic in shape and in nearly all cases the co-ordination number is six. It should be noted that either the cations or the anions form the close-packed structure. *Since cations are usually smaller than anions, the latter (anions) form the close-packed structure and cations are fixed in interstitial sites.* Each cation has an identical environment which is different from the identical environment of anions. **Every ion in a crystal interacts with every other ion so that the whole crystal can be considered as a single giant structure.** In sodium chloride crystal, each sodium ion is surrounded by six chloride ions and in turn each chloride ion is surrounded by six sodium ions. Sodium chloride lattice has a cubic structure as shown in Fig. 1.20.

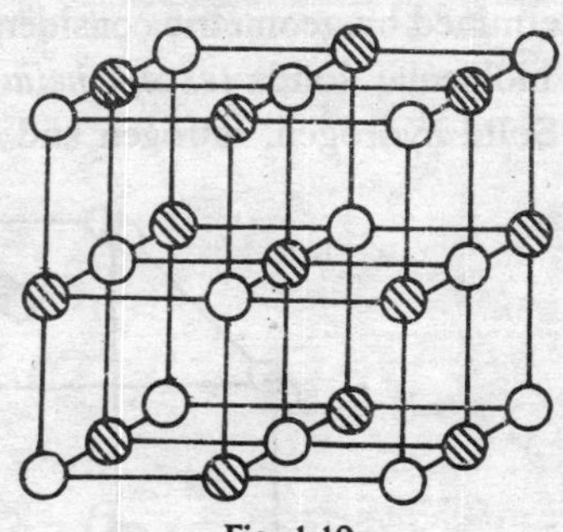

Fig. 1.19

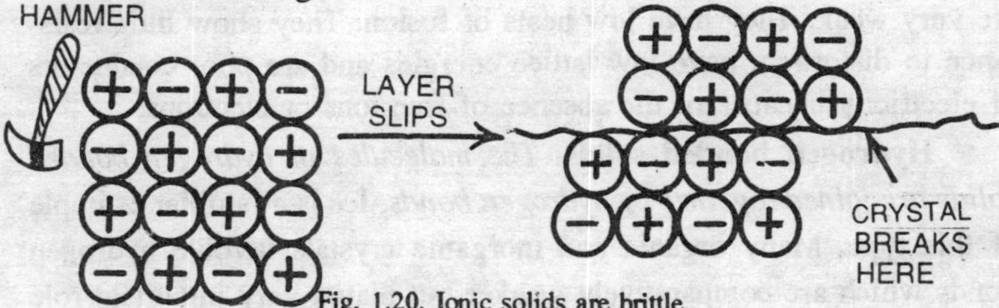

Fig. 1.20. Ionic solids are brittle..

Characteristics of ionic solids. Ionic solids have high melting and high boiling points because of the stronger attractive forces acting between oppositely charged ions. Ionic solids are hard and brittle since distortion brings ions of the same charge close together and repulsive forces then cause the crystal to fracture. Ionic sides are poor conductors of electricity because the ions are fixed and the applied electric field cannot make them move. However, they conduct electricity in molten state in which ions are free to migrate. Ionic lattices have high lattice energies.

(b) Molecular (van der Waals) solids. *The atoms or molecules in a molecular solid are held together by weak van der Waals forces.* Since van der Waals forces are non-directional, structure of the crystal is determined by geometric considerations only. The noble gases which form molecular solids *(except helium)* have cubic close-packed structure. Solid hydrogen, nitrogen and oxygen crystallise with hexagonal close packed structure. Other examples of this type of solids are carbon dioxide, iodine and sugar.

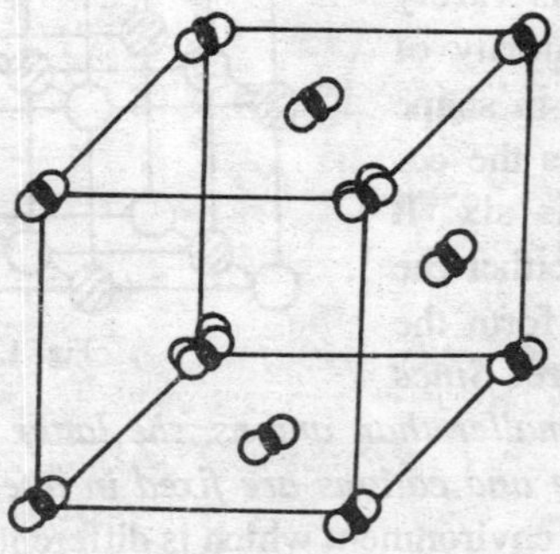

Fig. 1.21. Molecular solid of CO_2.

Characteristics of molecular solids. Molecular solids have low melting points and low boiling points because the intermolecular forces are very weak. They have low heats of fusion. They show little resistance to distortion, have low lattice energies and are poor conductors of electricity because of the absence of free ions or electrons.

Hydrogen bonded solids. *The molecules in hydrogen bonded solids are joined together by hydrogen bonds.* Ice is a familiar example of this type. Many organic and inorganic crystals involve hydrogen bonds which are comparatively weaker but play a very important role in determining the structures of substances such as proteins and polynucleotides.

For example, iodine crystallises with an orthorhombic lattice. At each lattice point is situated a diatomic molecule of iodine. The interatomic distance in the iodine molecules is 2.70 Å, but the distance of closest approach of two iodine molecule is 3.54 Å. This indicates the weakness of the van der Wall's forces compared to the strength of the covalent bond itself.

Other examples of molecular crystals are solid CO_2 , wax, sulphur and ice.

The molecular crystals have the following characteristics:

(*i*) They have small binding energy,

(*ii*) They have low melting and boiling points,

(*iii*) They are poor electrical conductors in solid, liquid as well as in dissolved state because these do not have ions or charged particles.

(*iv*) They have low heat of fusion,

(*v*) They are poor thermal conductors,

(*vi*) They have low heat of evaporation in comparison to the value for ionic or covalent crystals. Therefore, they are more volatile.

(*vii*) As the forces binding the molecules together in molecular crystals are weak, these are soft, easily compressible and easily distortable.

(c) Covalent solids. The atoms in a covalent solid are held together by covalent bonds. The geometry of such crystals is determined by well defined covalent bonds. A covalent bond is always directional and those covalent solids in which covalent bonding continues indefinitely in two or three dimensions are called *macromolecular crystals.* With the exception of graphite, they are poor conductors of electricity. This indicates that ions are not present in these solids and electrons are not free to move.

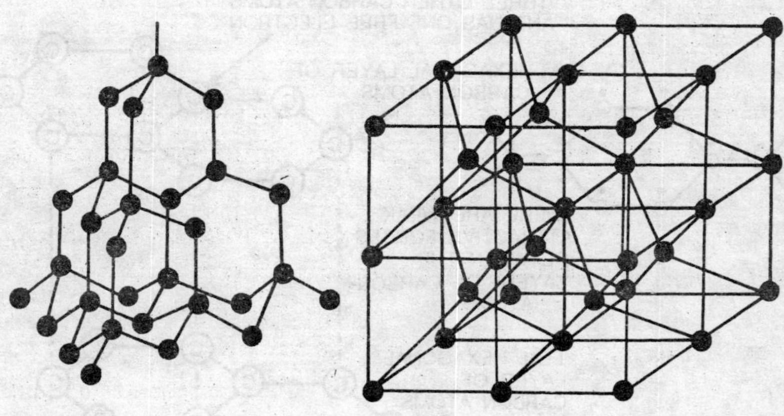

Fig. 1.22. Structure of diamond.

(i)* Three dimensional covalent solids.** An example of three dimensional net-work is the most familiar ***diamond crystal. Each carbon atom is joined to four other carbon atoms by sp^3 hybrid carbon-carbon bonds. All carbon atoms are tetrahedrally arranged because sp^3 hybrid orbitals lie at the corners of a regular tetrahedron. Diamond is a very hard substance which is used as an abrasive because of its extreme hardness. If attempts are made to bend diamond, it fractures in specific directions called cleavage planes. Diamond is a poor conductor to electricity because it does not contain free electrons. Silicon carbide has structure similar to that of diamond in which a silicon atom is surrounded tetrahedrally by four carbon atoms.

***(ii)* Two dimensional covalent solids.** Graphite is an example of two dimensional covalent solid. *Graphite consists of carbon atoms joined together by weak van der Waals forces.* In each layer, a carbon atom is joined to two other carbon atoms by sp^2 hybrid carbon-carbon bonds. The remaining π electrons are delocalized over the entire layer. The mobility of π electrons make graphite a good conductor of electricity. Graphite is a soft solid which is used as a solid lubricant because its layers can slide over one another due to the weak van der Waals forces between them.

***(iii)* One dimensional covalent solids.** A few non-metals in which covalent bonding is limited to a few atoms are also known. The

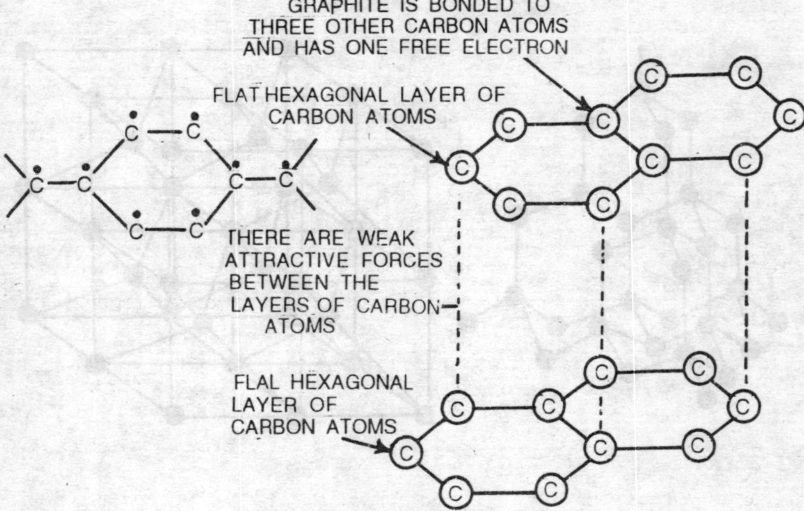

Fig. 1.23. Structure of graphite.

valence of such non-metals is two. They form chains ***(e.g., monoclinic sulphur spiral chains)*** **or rings** ***e.g., rhombic sulphur-puckered eight membered rings).*** **The chains or rings are held together by weak van der Waals forces.**

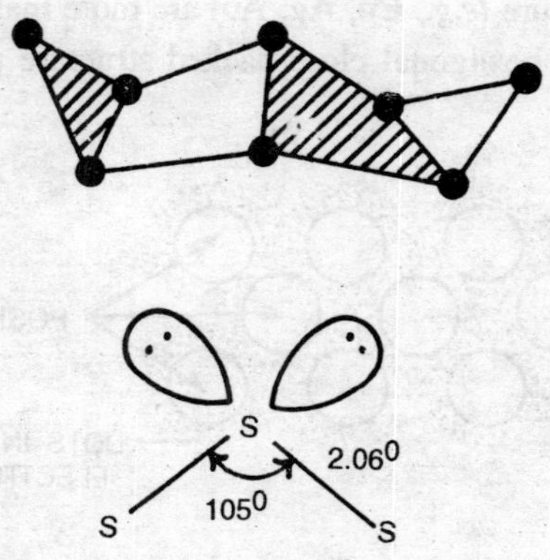

Fig. 1.24. The structure of rhombic sulphur.

Characteristics of covalent solids. Covalent solids are generally hard, brittle and have high melting points. With exception of graphite, they are poor conductors of electricity. This indicates that ions are not present in these solids and electrons are not free to move.

(d) Metallic solids. Solid metals have crystalline structure. Their appearance is not crystalline but all solid metals are in fact crystalline when viewed under magnification. To a good approximation, an atom of a metal is assumed to be a sphere. **The bonding in metals is non-directional and the atoms are held together by attractive forces.** A very simple picture of metallic structure is that the metallic cations are fixed in space in a regular pattern and free mobile electrons surround them. Metals have close-packed structures. Most of the metals have hexagonal or cubic close packed structures in which the co-ordination number is twelve, *i.e.,* each atom is surrounded by twelve other atoms. Some metals have less packed structure in which the co-ordination number is eight. A few metals have structures different from those described above. For example, indium has a tetragonal structure.

Characteristics of metallic solids. Metals are usually hard, brittle, have high melting and boiling points and are good conductors of heat and electricity. Presence of free electrons makes metals good electrical and heat conductors. Electrical conductivity of metals decreases as temperature is raised. Metals are malleable and ductile because planes of atoms can glide over one another. Metals having cubic close-packed structure (*e.g.*, Cu, Ag, Au) are more malleable and ductile than those having hexagonal close-packed structure (*e.g.*, Al).

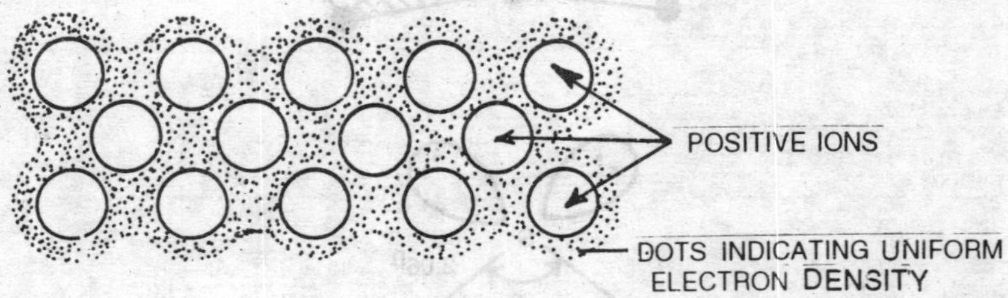

Fig. 1.25. Structure of metallic crystal.

Polymorphism. A few metals exist in more than one crystalline form and are then said to exhibit *polymorphism.* Iron has a body-centred cubic structure at normal temperature but a face-centred cubic structure at elevated temperature.

Heats of vaporisation and fusion of some typical crystals are given in the table 1.6:

Table 1.6.

Substance	*Heat of Vaporisation (kJ/mole)*	*Heat of Fusion (kJ/mole)*
Ionic crystals Sodium chloride	170.75	28.45
Molecular crystals Ammonia (solid)	23.35	5.65
Covalent crystals Graphite	718.43	—
Metallic crystals Copper	304.59	13.016

Table 1.5. Distinction between Various Types of Solids

Property	*Ionic*	*Metallic*	*Covalent*	*Molecular*
1	*2*	*3*	*4*	*5*
1. Examples	NaCl, LiF, $CaCl_2$, ZnS	Metals and alloys, Cu, Fe, Au, etc.	Diamond, graphite rhombic sulphur, quartz	Iodine, ice, sugar, noble gases
2. Constituent particles	A regular arrangement of cations and anions	Fixed metallic cations surrounded by free mobile electrons	Covalently bonded atoms	Molecules or atoms held together by weak van der Waals forces or Hydrogen bonds
3. Binding forces	Strong electrostatic attraction between ions	Electrostatic forces between cations and free electrons	Strong covalent bonds	Weak van der Waals forces or Hydrogen bonds
4. Hardness	Hard	Low as well as high hardness	Very hard Graphite is soft	Very low
5. Brittleness	Brittle	Low	Intermediate	Low
6. MP's	High	Intermediate as well as high	Very high	Low
7. Electrical conductivity	Bad conductors, conduct in fused state and solutions	Good conductors	Bad conductors Graphite is an exception	Bad conductors

	1	2	3	4	5
8.	Solubility in polar solvents	Soluble	Insoluble	Insoluble	Soluble as well as insoluble
9.	Solubility in non-polar solvents	Insoluble	Insoluble	Usually soluble	Soluble as well as insoluble
10.	Latent heat of fusion	High	High	High	Low

2

The Origin and Growth of Crystals

2.1. Introduction

Crystals appear in the course of *phase transformations, i.e.,* when a substance changes from one state to another.

The following are the basic conditions under which the formation of crystalline matter takes place:

(1) a change from the liquid phase to the solid-crystallization occurs from a melt or a solution;

(2) a change from the gaseous phase to the solid-crystallization occurs by sublimation;

(3) a change from one solid phase to another, accompanied by alteration of the shape of the crystal structure; this phenomenon is called *recrystallization;* it should not be confused with repeated crystallization, when mother liquor is poured off each time and the crystals are dissolved in a fresh portion of solvent.

The three cases of formation of crystalline matter are not equally widespread; the first case is observed much more frequently than the second or the third.

1. Crystallization from a melt or a solution often occurs on a very wide scale both under natural conditions and in industry, as *e.g.* the formation of massive crystalline rocks during solidification of a magma, deposition of thick salt beds at the bottom of salt lakes and periodically drying bays, solidification of molten metals, manufacture of various products at chemical plants, etc.

Deep-seated massive crystalline rocks are formed in three stages of crystallization: primary, principal, and residual.

During the *primary stage,* when the melt is still very hot, individual crystals are formed of those chemical compounds, *i.e.,* minerals, which have the highest melting point. The crystals can grow freely developing their particular forms. Such freely developed crystals are called *idiomorphic*.

As the melt cools, the number of substances crystallizing from the magma progressively increases, and the process of crystallization becomes more intense. The second, *principal stage of crystallization* begins. Individual crystals cannot develop freely because their growth is hindered by adjacent crystals of the same or other minerals: a struggle develops for free ions and free space. The boundaries of individual crystals often depend on the shape of the neighbouring or earlier formed crystals which prevent the free growth of later crystals.

During the last, *residual state* the change from the liquid to the solid phase is forced by a considerable drop of temperature. Stray ions fill the remaining vacant places, sometimes forming glass, *i.e.*, amorphous, not crystalline, substances.

When magma solidifies at or near the surface, cooling proceeds much more rapidly, and the whole process is considerably speeded up. In this case a larger quantity of material crystallizes during the last stage, thus forming volcanic glass.

2. During sublimation crystals are formed directly from vapour, by-passing the liquid phase. The specific feature of the crystals thus formed is usually their small size and sometimes skeletal form.

Sublimation occurs naturally in so-called *dry fissures,* on the walls of which various minerals are deposited from the rising gases as a result of their cooling. We can often observe this process in winter, when hoar-frost and various ice patterns are formed on window panes. Coloured haloes around the sun, and on frosty nights around the moon or some bright stars, are due to the refraction of rays in minute ice crystals formed in the atmosphere from water vapour.

In the chemical industry sublimation is used for separation and purification of many products (*e.g.*, iodine, naphthalene, camphor). In laboratories the results of sublimation appear as white coatings on reagent jars and glass panes of exhaust cabinets. Sublimates of different colours on charcoal or in test tubes serve in analytical chemistry, zinc, mercury, etc.).

3. The least studied process of the formation of crystals is recrystallization in the solid state. It consists of enlarging of tiny crystals of various aggregates under the action of deformation forces, periodic stresses, and temperature variations. For instance, pressure sometimes

causes chalk (submicroscopic crystals of calcium carbonate) to change into distinctly crystalline marble. Fine-grained iron of some parts of machinery becomes coarse-grained as a result of constant jolting and directed impacts. The process of recrystallization can be used for obtaining large single crystals of metal.

What is the initial stage of crystallization like? We know that in gases and liquids the molecules are perpetually in motion and that in gases molecules move much more rapidly than in liquids. If we accept the theory of intermolecular forces of attraction and repulsion, we can picture molecules graphically as dipoles with positive and negative charges concentrated at definite points (Fig. 2.1). As the temperature falls, the speed of molecular movement is reduced. With the transition to the solid phase the particles may be either arranged at random with respect to each other or may be mutually oriented. In the first instance we shall have an amorphous solid phase which can be regarded as a supercooled liquid. In the second instance the positive and negative regions of the molecules may become the primary nuclei of a crystal lattice upon assuming a stable position (see Fig. 2.1).

A nucleus is a microscopically minute crystal which is in a state of complete equilibrium with a supersaturated solution.

A strongly supersaturated (labile, *i.e.*, unstable) solution in which crystallization has started appears to be a mass of randomly moving particles (molecules, ions) and crystal nuclei (submicrons). The nuclei may be unidimensional, two-dimensional, and three-dimensional tiny crystals of different shapes and sizes. Some nuclei grow, others dissolve, but on the whole growth is predominant. Nucleus growth results either from adherence of individual molecules to a crystal, or from cohesion of nuclei, mainly two-dimensional ones or lamellate three-dimensional nuclei formed from two-dimensional ones. During their growth crystals tend to assume an equilibrium form; but ways of attaining an equilibrium form are not always the shortest possible, therefore initial crystal forms (still in the process of growth) differ considerably from forms which have attained equilibrium. If the total excess of substance in a supersaturated solution has crystallized before the crystals have assumed an equilibrium form, continued change in this form can proceed only at a very slow pace. Therefore, practically

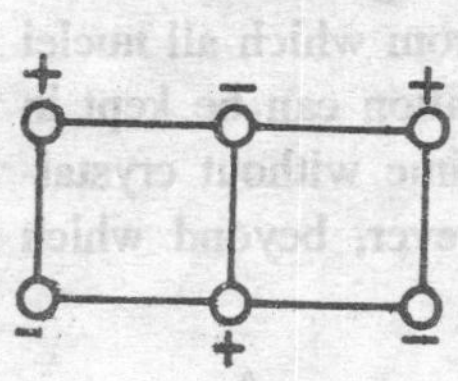

Fig. 2.1. Arrangement of molecules in nuclei

all crystals, irrespective or their form, will have attained equilibrium with the saturated solution. A slightly supersaturated (metastable) solution has no crystal ultramicrons, and crystal growth in it is possible only on introduction of nuclei from outside.

All liquids are capable of becoming supercooled to a greater or lesser extent. For instance, glycerine and melted salol can be cooled to a temperature much below their crystallization point without changing to the solid state.

When supercooling occurs, the curve assumes a different appearance (Fig. 2.2). It was obtained for a substance normally crystallizing at 23°C. Supercooling was carried down to 9°C (23° − 9° = 14°C). Crystallization thus started at + 14°, but the exothermal reaction raised the temperature to + 23°. In cases of considerable supercooling (dotted part of the curve) the temperature may not even rise to the normal level.

Viscous liquids, in which the molecules move much more slowly than in labile liquids, often solidify without crystallization on sufficiently rapid cooling, *i.e.*, they acquire an amorphous structure.

In the majority of cases the nucleation and growth of crystals occur at a lowering in temperature of the melt and the accompanying supersaturation of the solution. Initial crystallization may be artificially induced (via nucleation) or spontaneous. In nucleation, tiny crystals of the crystallizing substance itself are best, although dust particles with a structure closely resembling that of the crystallizing substance often prove suitable enough. Supersaturated solutions from which all nuclei have been carefully removed during their preparation can be kept in hermetically sealed vessels for long periods of time without crystallization. Supersaturation has certain limits, however, beyond which spontaneous crystallization is inevitable.

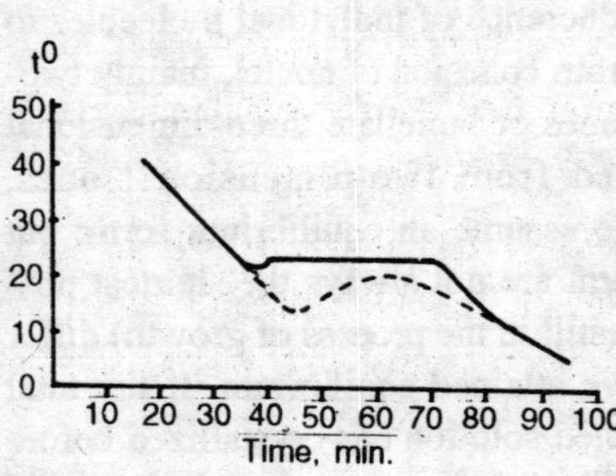

Fig. 2.2. Cooling curve for a substance crystallizing from a supercooled solution.

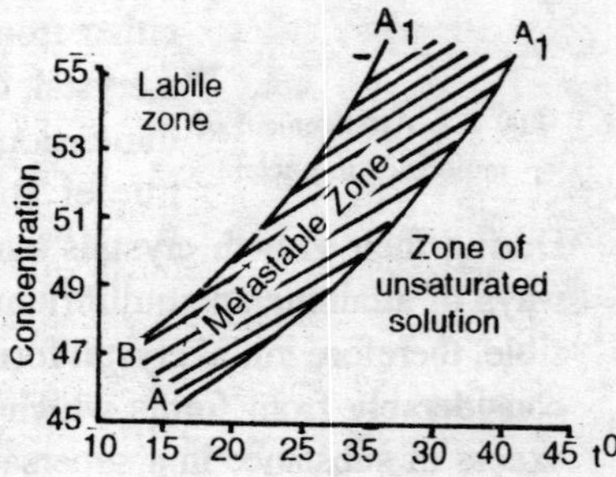

Fig. 2.3. Three zones in a solution.

The process which takes place in such a case can be explained as follows. Above the solubility curve AA_1 (Fig. 2.3) there is another curve BB_1 which is called a *supersaturation curve*. To the right of the AA_1 curve is the zone of unsaturated solutions. To the left of the BB_1 curve is the zone of *labile solutions* in which crystallization begins spontaneously. Between the two curves is located the metastable[1] *zone*; the ratio of concentration and temperature for forced crystallization is dependent on the boundaries of this zone.

Changes in crystals with time occur mainly on their surface. Crystal growth results from the deposition of the compound of which it is composed on its surface, therefore it may be said that every particle of a crystal formed was at some time in its existence a part of that surface. Hence the shape of a crystal is its present, while the internal structure (and texture) is its past.

A natural crystal, as a mineralogical phenomenon, is a heterogeneous entity consisting of closely interconnected parts.

The principal laws of crystal growth can be established through a study of sections of natural or artificial crystals. If in the course of growth some factor affecting the homogeneity or chemical composition of the crystal *should vary*, the latter will acquire a so-called *zonar structure* (Fig. 2.4), which frequently occurs in minerals and can be observed sometimes in smoky quartz, where weak and strong colouring alternate in layers; or in crystals of fluorite (CaF_2) where

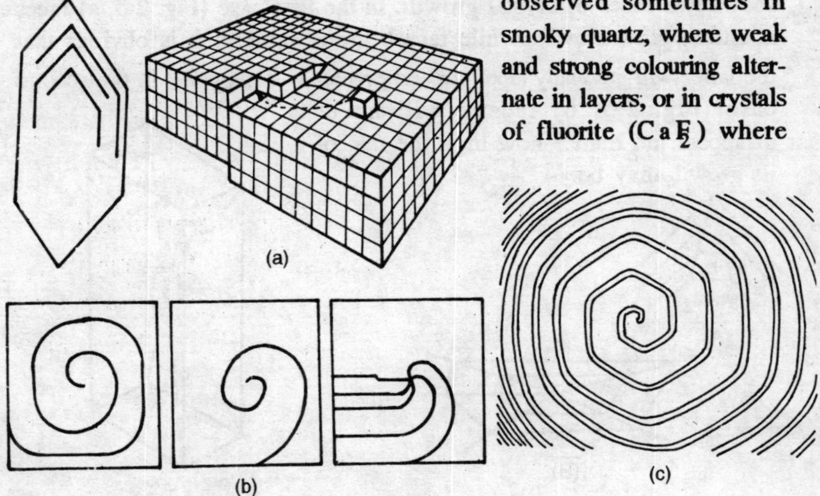

Fig. 2.4. Zonar structure of a crystal.

1. This word "metastable" is of Greek origin and means nearly stable.

colourless layers often alternate with those tinted bright violet or green.

Examination of crystals with a zonar structure will show that *during growth the faces of a crystal are translated parallel to themselves.* The measurement of the thickness of different layers in crystals with a zonar structure shows that different faces grow at different rates (by the rate of growth is meant the increase in thickness per unit time). The growth of a crystal face does not occur as a deposition of plane layers, for new material particles are deposited in a spiral (Fig. 2.4, a,b). Traces of such deposition can be observed on the faces of crystals which have ceased growing, *i.e.*, which have been removed from the solution (Fig. 2.4, c).

Knowing the shape and the dimensions of a crystal at a particular moment, as well as the rate of growth of its faces, we can predict what its shape will be at any point in the future (taking the rate of growth to be constant). To do this it will suffice to draw lines normal to the crystal faces and proportional to the rate of growth, and to draw planes perpendicular to these lines until the planes intersect. The relative dimensions of faces will alter in this process: some faces will grow larger, and others smaller. In the majority of cases those faces will increase which have the lowest rate of growth; the rapidly growing faces gradually become smaller and may finally disappear completely. This rule holds true only under particular geometrical conditions. Fig. 2.5 shows two cases of crystal growth. In the first case (Fig. 2.5, a) faces a_1 and a_2 grow slowly, while face b grows rapidly; it is obvious that the latter will gradually taper out and eventually disappear. In the second case (Fig. 2.5, b) face k will never disappear, no matter how high the rate of its growth may be.

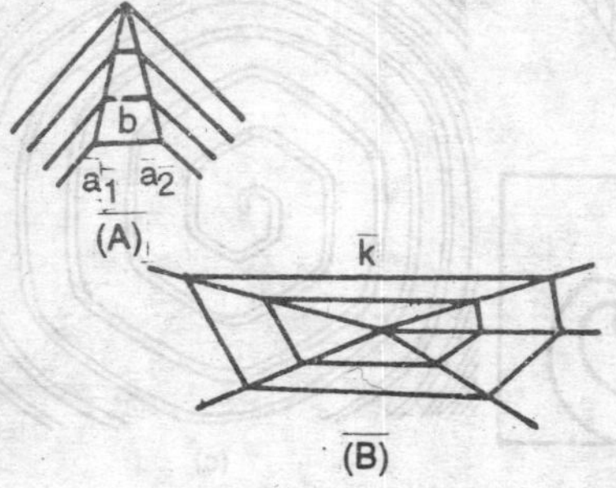

Fig. 2.5. Growth of faces.

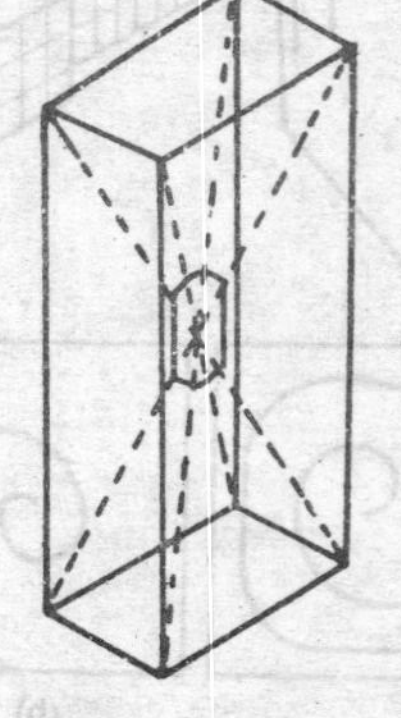

Fig. 2.6. Pyramids of growth.

(a-) face b tapersont; (b-) face k does not taper out.

When a crystal grows uniformly, its coigns progress along straight lines, the edges progress along planes, while the loci of faces form pyramids of growth. All these loci converge to a single point—the initial nucleus of crystallization (Fig. 2.6). Actually such an orderly arrangement of pyramids of growth is exceptionally rare. Usually the loci of development are broken and curved because of non-uniform growth. Crystals are never perfectly uniform. It can hardly be imagined that the solute could be deposited from a solution on a growing face in a perfectly plane layer. It is much more probable that this deposition will occur through adhesion of molecular aggregates to a face, the aggregates inevitable differing from each other in size, though ever so slightly. Let us imagine an ideal face. When the first layer is being deposited, the molecular aggregates are arranged parallel to each other, but with free space between them. In the next layers incomplete parallelism begins to tell, as a result of the varying thickness of the aggregates deposited earlier. As the latter decrease in size, the structure of the crystal will approach the ideal pattern. In a considerably supersaturated solution crystals grow almost exclusively as a result of adhesion of spontaneously developing nuclei. The molecular aggregates are naturally large, and the spaces between them are filled with mother liquor and foreign matter. The resulting crystal is turbid and has low mechanical strength. In a slightly supersaturated solution conditions are more favourable for obtaining crystals of nearly ideal structure, but here too the degree of uniformity is inversely proportional to the rate of growth.

The following factors have the greatest effect on the shape of the growing crystal.

1. Concentration Currents. The portions of a supersaturated solution directly adjoining a crystal give up their excess of solute to the crystal. As the concentration of these portions of the solution thus decreases, the solution becomes lighter and rises up. Its place is taken by heavier portions, and the cycle is repeated. As a resusIt, a zone of weaker concentration is built up above the growing crystal. The currents that develop in the process are called *concentration* currents. The larger the growing crystal, the stronger are the currents (Fig. 2.7).

If the crystal lies at the bottom of a vessel, its side faces have the most favourable conditions for growth. The lower parts of the faces are surrounded with more supersaturated solution than are the upper parts, this resulting sometimes in the appearance of steps and furrows

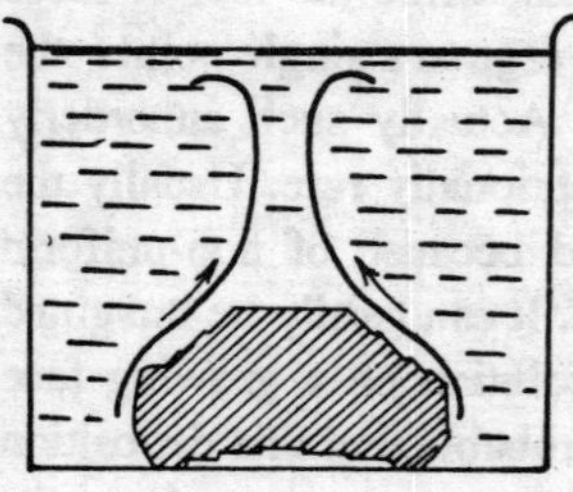
Fig. 2.7. Concentration currents.

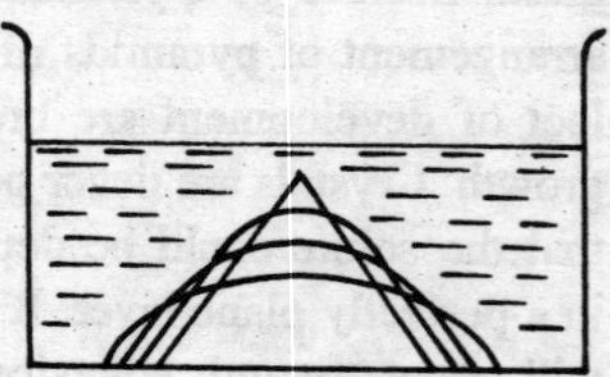
Fig. 2.8. Effect of alternate increase and decrease of temperature on a crystal growing in solution.

on the lower portions. The faces inclined at obtuse angles to the bottom of the vessel grow more rapidly than those forming acute angles. The top face, which is the slowest growing one, is often very well formed (*i.e.*, it is plane and smooth), if there have been no interfering eddies. The bottom face either does not grow at all, or grows only along the edges because of supersaturation diffusion; in this case a stepped indentation is formed in the face (see Fig. 2.7).

If the crystal is suspended in the solution, its bottom face is the one best supplied, and grows more rapidly than the other faces, and the crystal becomes elongated downwards.

The concentration currents strongly influence the shape of the growing crystal: if it is at the bottom of a vessel, the currents tend to make it flat; if it is suspended, they tend to elongate it.

The effect of the concentration currents can be eliminated by rotating the crystal about a horizontal axis or by shaking the crystallizer flask.

2. Temperature Variations. This factor determines to a large extent the solubility of most substances and the shape of a growing crystal.

Let us imagine a crystal in a saturated solution at the bottom of a flat vessel (Fig. 2.8). When the temperature rises, the solution becomes unsaturated and the crystal begins to dissolve. The more concentrated solution thus formed flows down, preventing the lower part of the crystal from dissolving any further and facilitating the dissolution of its upper part. The solution becomes laminary in the process. When the temperature is lowered, the crystal grows mainly in its lower part, as the solution there is more concentrated, while the concentration currents cause the displacement of the solution. This cycle is repeated at each rise and drop in temperature. The crystal becomes flatter and flatter until it is a plate of zonar structure occupying the whole bottom of the vessel. A crystal suspended by a thread will under these conditions soon drop to the bottom, as its top part, to which the thread is affixed, dissolves. If two crystals are placed one above the other, the lower crystal will grow at expense of the upper.

It should be noted at temperature variations often cause faces to appear which would never be formed under normal conditions of growth.

3. Degree of Supersaturation of Solution. This factor affects the homogeneity of crystals, which are less homogeneous when formed in strongly supersaturated solutions, and more homogeneous in slightly supersaturated solutions. The degree of supersaturation has, moreover, a considerable effect on the shape and the number of crystal faces.

4. Impurities in the Solution. Impurities often strongly affect the crystal shape. A classical example is the introduction of carbamide into a solution of sodium chloride. Without this addition sodium chloride crystallizes in cubes, with it—in octahedra.

5. Viscosity of the Solution. If the viscosity is high enough, it will prevent the formation of concentration currents. Then the crystal will be able to grow only through diffusion of supersaturation, which has a peculiar effect on the shape of the growing crystal.

To understand what goes on in this case, let us imagine that the crystal has only one infinitely large face. The supply of solute must come along lines which are normal to this face and which we shall call lines of *diffusion current.* These lines are uniformly spaced throughout the face. Actually the lines will be curved near the coigns and edges of the crystal (Fig. 2.9, a) as the faces are supplied from portions 1, 2, 3 and 4 (Fig. 2.9, b), while the edges and coigns have additonal feeding portions *k, l, m* and *n*. Therefore, the crystal grows more intensely in the *k, l, m, n* portions where correspondingly thicker parts appear (Fig. 2.9, c). These thicker parts unbalance still further the distribution of feed to the individual parts of the crystal, and therefore it grows mainly in coigns, developing so-called *skeletal forms* (Fig. 2.10). These forms can also develop in a strongly labile medium which for one reason or another is not stirred. Sometimes skeletal forms develop on sublimation.

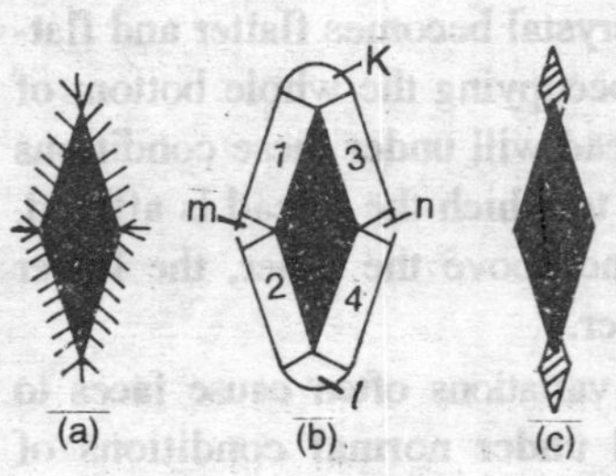

Fig. 2.9. Development of skeletal forms in crystallization from a viscous solution.

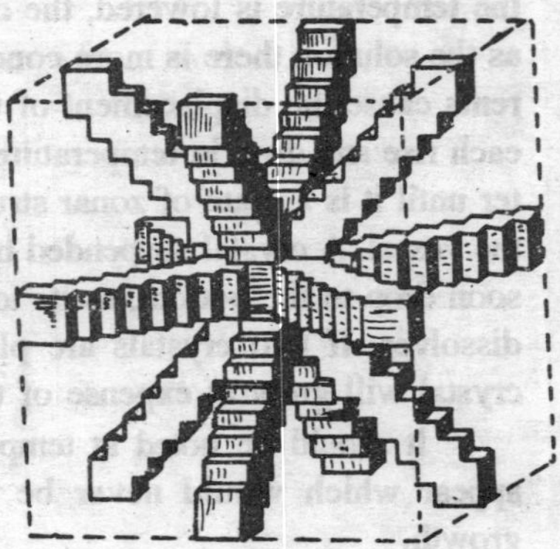

Fig. 2.10. Skeletal form of a crystal of metallic copper.

2.2. Equilibrium Form of Crystals

Formation of an interface at constant temperature, pressure, and quantity of substances making up the phases involves some expenditure of energy. The work done in forming a unit interface is called surface tension and designated by σ.

As the rates of growth of different faces of the same crystal are different, the faces have different surface tension.

The *equilibrium crystal form* is a form which at a given volume of the crystal corresponds to the minimum surface energy expended in its formation. This form should satisfy the following two requirements:

$$E = \sigma_1 s_1 + \sigma_2 s_2 + \ \ldots\ldots = \min$$

$$V = \text{const.},$$

where s_1, s_2 are areas of faces with surface tension $\sigma_1, \sigma_2 \ldots\ldots$;

E is the total surface energy of a polyhedron;

V is the volume of a polyhedron.

Gibbs (1878) put forward the theory that a crystal tends to assume an equilibrium form during growth. To acquire the most stable form, the one we call an equilibrium form, a crystal must be repeatedly thrown off balance by the introduction into it of small amounts of energy in the form of heat. Thus, the crystal will have an opportunity of returning during subsequent cooling that quantity of energy which it can release under given conditions. It has been shown experimentally that a ball made from an alum crystal and placed in a saturated solution becomes an octahedron of the same weight in the course of several months under influence of temperature variations (*i.e.*, as a result of repeated growth and dissolution). This octahedron is obviously the equilibrium form of an alum crystal in a saturated aqueous solution.

It is evident from the above that crystal forms—both natural and artificial—usually encountered in practice, are not equilibrium forms at all.

We may visualize three types of crystalline formation: *unidimensional, two-dimensional,* and *three-dimensional.* In formations of the first type (molecule, atom, ion, a group of atoms) crystal particles form chains, in formations of the second type-polygonal layers, and in formations of the third type—polyhedra.

2.3. Release of Energy in Crystallization

We have already seen that crystallization is accompanied by a release of energy, usually in the form of heat. If the quantity of energy released in forming a unit weight of each of the three types of crystalline formation be designated as a_1, a_2, a_3, respectively, we shall have the relation $a_1 < a_2 < a_3$, since there are more unutilized bonds on a crystal surface in unidimensional and two-dimensional formations, than in three-dimensional ones.

The above is true of large crystals. In very small crystals the quantity of heat released greatly depends on the position of each individual particle. For instance, if a particle adheres to a crystal with one of its faces, the quantity of heat released will be smaller than if the particle adhered to the crystal with two or three faces. There is no direct relationship, however, between the quantity of heat released and the number of adhering faces, becasue the place at which the particle adheres—in the middle of a face, at the edge, or at the coign of a polyhedron—is also important. Theoretically cases are possible in which less energy is released in the formation of very small three-dimensional crystals consisting of a few molecules, than in the continued growth of a large two-dimensional crystal.

In the course of slow, undisturbed growth particles are deposited on a crystal mainly in a manner producing the greatest release of energy.

If a crystal is at the stage of growth shown in Fig. 2.11 greatest quantity of energy will be released when a particle adheres to a three-faced corner *(1),* a smaller quantity, when it settles in a two-faced corner *(2),* and still smaller, when it settles on a face *(3).* This explains why a second layer cannot be formed on a face until the formation of the first layer has been completed. This does not mean that the second layer cannot begin developing before the first one is completed; it can begin growing, but the conditions will not favour it. Therefore, in the case of two-dimensional crystal its growth is more easily continued in the form of a plate than by transformation into a three-dimensional crystal through the addition of a second layer.

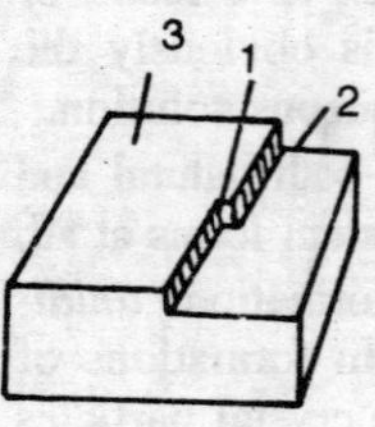

Fig. 2.11. Three cases of implanting a particle during growth.

2.4. Characteristic Features of the Dissolution of Crystals

A crystal placed in a pure solvent or in an unsaturated solution begins to dissolve. The coigns and edges dissolve quicker than the faces, and the crystal becomes somewhat fused in form. This is explained by the fact that the particles released when the outermost portions of faces are dissolving have greater freedom of movement than those from the centre, they break away more easily from the crystal, and provide access for the solvent to the underlying layers.

Observations have shown that the process of dissolving is characterized by the following phenomena:

1. While during crystal growth the most stable faces are the slowly growing ones, in dissolution those faces appear on a crystal that dissolve the fastest.
2. The rate of dissolution is a continuous function of the direction of the crystal face growth.
3. In the course of dissolution microscopic multifaceted pits are always fromed on crystal faces—these are the so-called *etch figures* or, in other words, *dissolution figures,* which are of considerable morphological significance. They are most easily obtained through the action of a weak solvent, and will either cover the face completely, making it appear dull, or they will form at only a few points. In the latter case the etch figures are usually fairly large and distinct. The opposite phenomena-*mounds of growth* appear much more rarely and are of an entirely different origin, since they are produced not only by growth, but also by periodic growth-dissolution cycles.
4. The most perfect surfaces are developed on those faces of a growing crystal which have the slowest growth rate. The surfaces with the slowest dissolution rate always appear full.

Crystals of some organic substances, when heated, turn into liquids that have the optical properties of uniaxial crystals. Such formations are called liquid crystals or crystalline liquids. About 250 such formations are known at present. Most of them are organic compounds of the aromatic series. They consist of molecules of a greatly elongated form. For instance, paraazoxyphenetol has this structure:

$$C_2H_5O-C_6H_4-N(O)=N-C_6H_4-OC_2H_5$$

A liquid crystalline substance is formed as a turbid liquid consisting of drops of homogeneous liquid crystals oriented in different directions. A good example of this is sodium oleate (soap). Structurally, liquid crystals can neither be classed with amorphous nor with crystalline substances.

2.5. Practical Methods of Obtaining Single Crystals from a Melt

Growing large individual crystals from a melt in the form of natural polyhedra is a difficult task, and therefore experiments are usually limited to obtaining single crystal blocks.

Three methods of growing single crystals from a melt are generally used: (1) the Kyropoulos method; (2) the Bridgman method (with a drawn-out tube); (3) the Verneuil method.

All these methods are based on the principle of selection. If, for instance, a number of small crystals start developing at the same time at the bottom of a vessel, in the course of further growth their number will decrease, and finally, only one crystal will "survive" and keep on growing. The reason for the selection is concealed in the geometry of crystal growth and in the initial orientation of crystals.

1. The *Kyropoulos device* (1926) consists of an electric resistance stove *A*, a platinum crucible *B* and a cooler *C* (Fig. 2.12, a). The cooler is a platinum test tube through which air can be blown by means of an inner copper tube. The cooler can be lowered mechanically into the crucible and very slowly withdrawn from it.

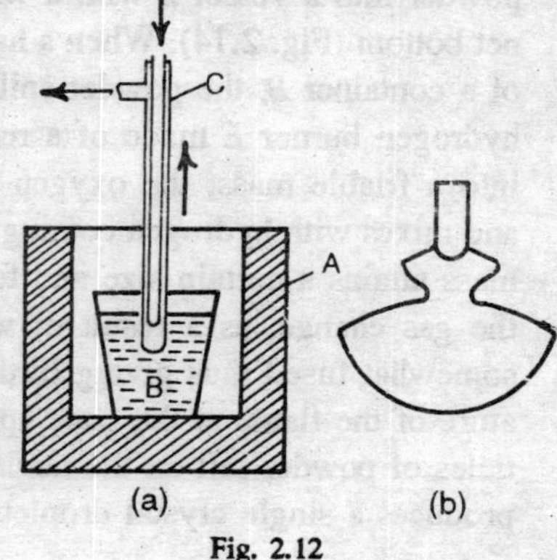

Fig. 2.12

When the substance selected for crystallization has been melted in the curcible and considerably superheated, the cooler is lowered into the melt and air is blown through it. Blowing is increased and the cooler is slowly lifted when a hemispherical growth of fine crystals has formed at its tip (Fig. 20, b). When the hemisphere comes into contact with the surface of the melt, only one of its numerous crystals usually survives and continues growing as the cooler is lifted higher.

This is an efficient method for growing fairly large crystals of sylvine (KCl) and of many other salts and pure metals.

2. Single crystals of fusible metals are obtained by the *drawn-out tube* method (Bridgman, 1925). A heat-resistant glass or quartz tube with the bottom end drawn out into a capillary is filled with a melt and suspended vertically in an electric tube furnace (Fig. 2.13). The metal is cooled by gradual lowering of the glass tube into another tube, of lower temperature. Crystallization begins in the capillary where the selection phenomenon takes place. Eventually the whole tube is filled by a single metal crystal.

The Bridgman method is excellent for obtaining single crystals of zinc (melting point 419°).

3. For heat-resistant substances, *e.g.*, for corundum (Al_2O_3), the *Verneuil method (1934)* is used. The substance to be crystallized is placed in the form of a very fine powder into a vessel *L* with a wire net bottom (Fig. 2.14). When a hammer *M* periodically strikes the wall of a container *B*, the powder spills along a tube *A* into the flame of a hydrogen burner *E* made of a refractory material where it is sintered into a friable mass; the oxygen supply comes also down the tube *A* and mixes with hydrogen coming in at *F*. As the cone *N* of the sintered mass attains a certain size and form, the conditions of combustion of the gas change, as a result of which the apex of the cone becomes somewhat fused thus giving birth to a nucleus. After that the temperature of the flame at the cone apex is so adjusted that the melted particles of powder fall on the nucleus which, steadily growing upwards, produces a single crystal droplet.

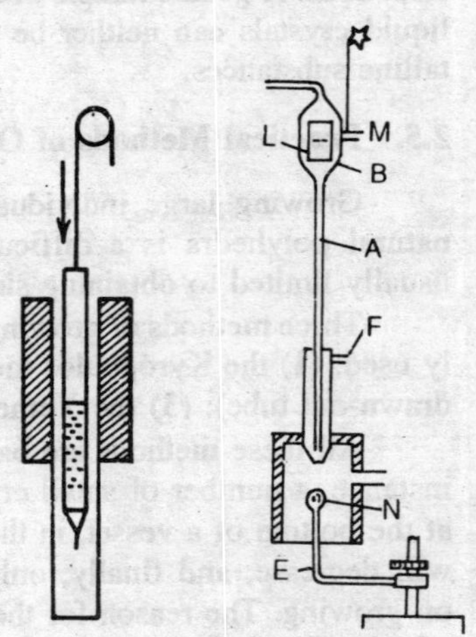

Fig. 2.13. Bridgman device

Fig. 2.14. Verneuil device

Here, as in other instances of single crystal growing described previously, only one of the many initially formed crystals survives.

S.K. Popov, a research worker at the Institute or Crystallography of the Academy of Sciences of the U.S.S.R., has designed a device for producing not droplets but thin single-crystal rods of cylindrical shape. This is of great practical importance, as crystals of red corundum are widely used in the watch-making industry as "stones". Insofar as these stones are of cylindrical shape, they can be easily obtained by sawing the rod across.

2.6 Crystallization from Solution

Crystallization from solution consists in the giving off of excess solute by the solvent. Hence crystals can be formed only from supersaturated solutions.

The *Coefficient of solubility* may be defined as that weight of substance which will dissolve in 100 g of solvent.

Supersaturation can be brought about in two ways: (1) by removal (evaporation) of the solvent; (2) by lowering the temperature, as this will result in a decrease of solubility for most substances.

The shape of the solubility curve will show which of the two methods should best be used in a given case.

Two types of solubility curves may be singled out (Fig. 2.15).

In one case (KNO_3) an increase in temperature of the solution results in a very sharp increase in solubility. In the case of NaCl, however, an increase in temperture from 0° C to 100° C results only in a very small increase in solubility—the quantity of substance which can be dissolved increases by only 5 g. Hence, to obtain some nitre crystals quickly it is necessary to dissolve the substance in boiling water and then to cool the solution. But this method is obviously impractical in the case of sodium chloride; to obtain its crystals the solution should be vigorously evaporated.

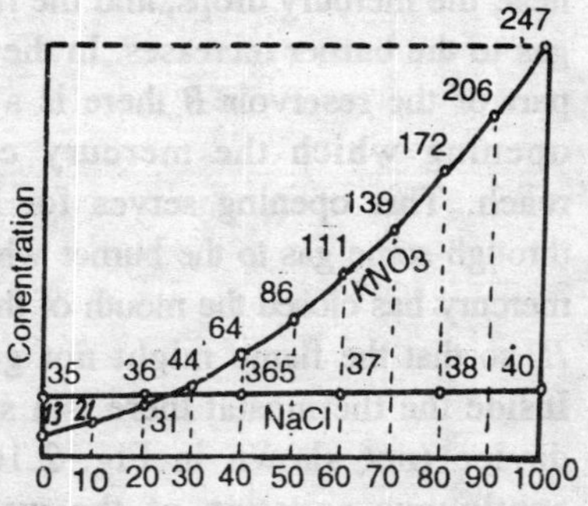

Fig. 2.15. Solubility curves for KNO3 and NaCl.

In the great majority of cases temperature variations strongly affect the solubility of substances, therefore crystallization has to be conducted in thermostats—devices with automatic temperature control. There are a number of thermostats of different types. Air is the medium of temperature control in some, while in others it is water. Crystallization is mostly conducted in water thermostats. They were devised by Professor G.V. Wulff and perfected by his student Professor A. V. Shubnikov. Heating can be done by electricity or gas.

The gas-heated water thermostat designed by G. V. Wulff (Fig. 2.16) consists of a metal tank several tens of litres in capacity, with thermally insulated walls. It is heated from below by one or more gas burners with fish-tail attachments to prevent back-firing when the flow of gas decreases considerably.

On the way from the mains to the burner the gas passes through a thermoregulator, which is designed as follows. The reservoir *M* (Fig. 2.16) is filled with a liquid which has a high coefficient of expansion, and therefore the section *M* becomes very sensitive to temperature variations. In Wulff thermostats the reservoir *M* of the thermoregulator is filled with toluene. A tube *p* is filled with mercury, the excess of which overflows into the reservoir *B*. Gas flows along the tube *H* and passes to the burner via the tube *B* (see arrows in Fig. 2.16). As temperature rises, the toluene expands, and the level of the mercury in the reservoir *B* rises, thus limiting the space through which the gas passes, whereupon the burner gives less heat, the mercury drops, and the flow of gas to the burner increases. In the upper part of the reservoir *B* there is a small opening which the mercury cannot reach. This opening serves for letting through some gas to the burner when the mercury has closed the mouth of the tube *H*, so that the flame might not go out. Inside the thermostat there is a special device (not shown in Fig. 2.16) for continuous agitation of the water in the vessel. Crystallizers are placed on special supports so that they are largely submerged. The water level must of course be below the walls of the crystallizers. The Soviet scientist A. V. Shubnikov has perfected Wulff's scheme in the following manner. Instead of a mechanical agitator he has introduced a pneumatic one

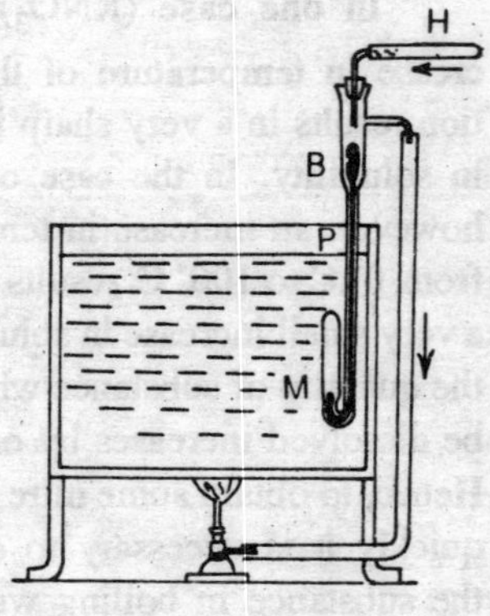

Fig. 2.16. Wulff thermostat.

(*LMN* in Fig. 2.17). A device OP has been added to maintain a constant level of water in the thermostat. The thermoregulator C has been made more convenient to use, and toluene has been replaced with mercury, since toluene vapours, penetrating between the mercury and the tube walls, attack the connecting rubber tubes. The Shubnikov thermostat has a capacity of about 1,000 litres. If well made, it will control temperature to within 0.01° C.

When crystals are grown by evaporation of a solvent, a crystallizer is placed in a hermetically sealed vessel containing a substance which absorbs the vapour of the solvent. Concentrated sulphuric acid is generally used in crystallization from aqueous solutions.

With small crystallizers it is convenient to use conventional desiccators (Fig. 2.18). The desiccator should be kept in a thermostat during crystallization, since even very small temperature variations cause defects in crystals. When the temperature rises, coigns and edges become fused, and at a subsequent drop in temperature vigorous growth begins in these spots, since the crystal tends in the first place to restore its multi-faceted form. Too rapid a growth results in the inclusion of small bubbles of solution, and often turbid portions appear in the crystal.

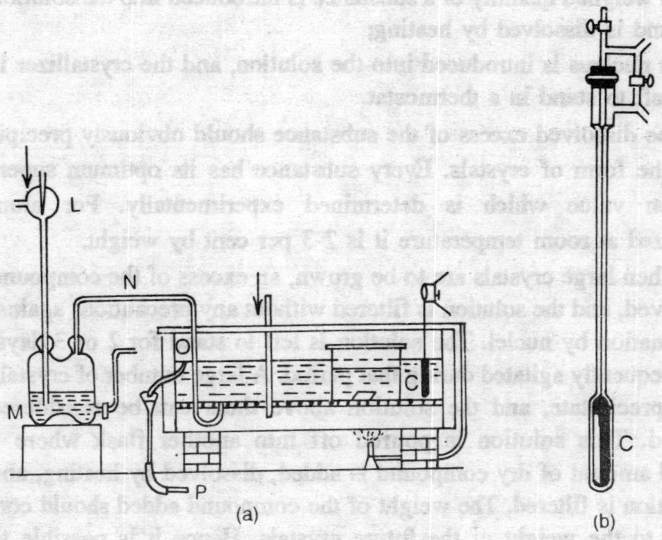

Fig. 2.17. Shubnikov thermostat:
a—general view of a thermostat: *b*—thermoregulator.

Some slightly heated saturated solution is poured into a crystallizer and a tiny crystal is introduced as a nucleus. These operations should be conducted in as sterile conditions as possible, to preclude

Fig. 2.18. Crystallization in a desiccator.

Fig. 2.19. Crystallization in a water thermostat.

contamination of the solution by dust-size particles of the crystallized substance.

Growing of crystals from solution by lowering the temperature is carried out in the following way:

(1) A solution is prepared, which is saturated for the temperature at which crystallization is to be conducted;

(2) a weighed quantity of a substance is introduced into the solution and is dissolved by heating;

(3) a nucleus is introduced into the solution, and the crystallizer is left to stand in a thermostat.

The dissolved excess of the substance should obviously precipitate in the form of crystals. Every substance has its optimum supersaturation value which is determined experimentally. For alum crystallized at room temperature it is 2-3 per cent by weight.

When large crystals are to be grown, an excess of the compound is dissolved, and the solution is filtered without any precautions against contamination by nuclei. The solution is left to stand for 2 or 3 days, and is frequently agitated during that period. A large number of crystals should precipitate, and the solution above them can be considered saturated. This solution is poured off into another flask where a weighed amount of dry compound is added, dissolved by heating, and the solution is filtered. The weight of the compound added should correspond to the weight of the future crystals. Hence it is possible to determine the volume of solution necessary for growing crystals of definite weight. A small well-formed crystal is placed in a still warm solution, the crystallizer is placed in a thermostat and sealed hydraulically (Fig. 2.19) to prevent contamination by nuclei from the atmosphere. If sufficient sterility has been observed, a single crystal of pre-calculated weight will grow in the vessel in 10-15 days.

2.7. Crystal Growth

Diffusion and convection are two important mass and heat transfer process governing the growth rate, morphology, chemical composition and homogeneity of crystals. C.T.R. Wilson (1900) and J. Frenkel (1932) independently developed the theory of intrinsic growth kinetics. On freezing, the crystal growth rate is controlled by the rates of atoms joining and leaving the interface. Above the equilibrium temperature, T_E, the crystal size diminishes while below it, it grows.

Techniques of Single Crystal Growth

The techniques of crystal growth may be divided into three categories: (a) growth from solutions, (b) growth from melts, and (c) growth from vapour.

(a) Growth from Solutions. The growth from solutions takes place in two stages: (i) nucleation and (ii) growth. Therefore, the first requirement is that of a suitable solvent which can dissolve the solute to an appreciable extent (ca. 20% solubility). This may be possible either under normal conditions (atmospheric pressure and room temperature) or high pressures/temperatures. The solvent may be water or melt of a metal.

Nucleation can be carried out from highly supersaturated solutions, so that it can occur spontaneously; but conditions should be such that the solution becomes metastable after a few new nuclei have formed. By metastable condition we mean a very small supersaturation, which prevents formation of many nuclei. The growth can then occur over the few 'seed' nuclei already formed. As the growth proceeds, the supersaturation decreases. In order to maintain a reasonable growth rate, the supersaturation has to be kept constant. This can be achieved by (*i*) isothermal evaporation of the solvent, (*ii*) lowering the temperature, (*iii*) adding solute or (*iv*) adding another solvent. Evaporation at a constant temperature is regulated by controlling the vents on the lid of the container and this procedure is suitable in cases when temperature dependence of solubility is small or when the solubility decreases with increasing temperature. When the solubility increase with increasing temperature, it is better to maintain supersaturation by decrease in temperature (~0.5° per day); this method is suitable when the growth is over around room temperature. For the production of seeds, it may be necessary to evaporate the solvent and decrease the temperature simultaneously. Fig. 2.20 depicts the experimental apparatus that can be employed. Sometimes the rotation of the growing crystal or stirring of the solution is done to equalize concentration gradient. Good crystals of alum can be grown from aqueous solutions because of the high solubility and structure similarity between the solvent and the (dissolved) solute. Some other examples are NaCl, KCl, $KNaC_4O_6 . 4H_2O$ (Rochelle salt) and $C_2H_4(NH_2)_4C_4H_6O_6$ (ethylenediamine tartrate).

(b) Hydrothermal growth. If the solubility of the solute in water is not high, growth from aqueous solutions becomes inconvenient. Since the solubility generally increases with increasing temperature, the temperature of operation may be increased beyond 100° C. This calls for employment of high pressure containers (autoclaves). Growth under conditions of high temperatures and high pressures is called hydroghermal growth and is employed for the preparation of some important materials like sapphire (Al_2O_3), yttrium iron garnet ($Y_3Fe_5O_{12}$) and β–quartz. We shall consider the last-mentioned material for illustration. β–Quartz is stable only upto 575° C. Therefore, it cannot be grown through high temperature methods like melt or vapour growth. On the other hand, it is not readily soluble is any solvent upto about 200° C. However, near 374° C—the critical temperature of water — it dissolves readily in pure water or an alkaline solution. Fig. 2.21 depicts an apparatus generally employed for the preparation of β–Quartz

(c) **Growth from Melt Solutions.** In the absence of proper solvents, growth may be carried out from melts of solids, which do not react with the solute. In order to grow materials with high melting points, some 'flux' materials may be added, which bring down the melting point. Several important semiconductors like gallium phosphide (GaP) and ferroelectrics like $BaTiO_3$ can be grown in this way.

Fig. 2.22 shows the liquids curve for the GaP growth from a melt of gallium containing phosphorus. The point A corresponds to the approximate composition 99.95 mole % Ga and 0.05 mole % P at 1300° C.

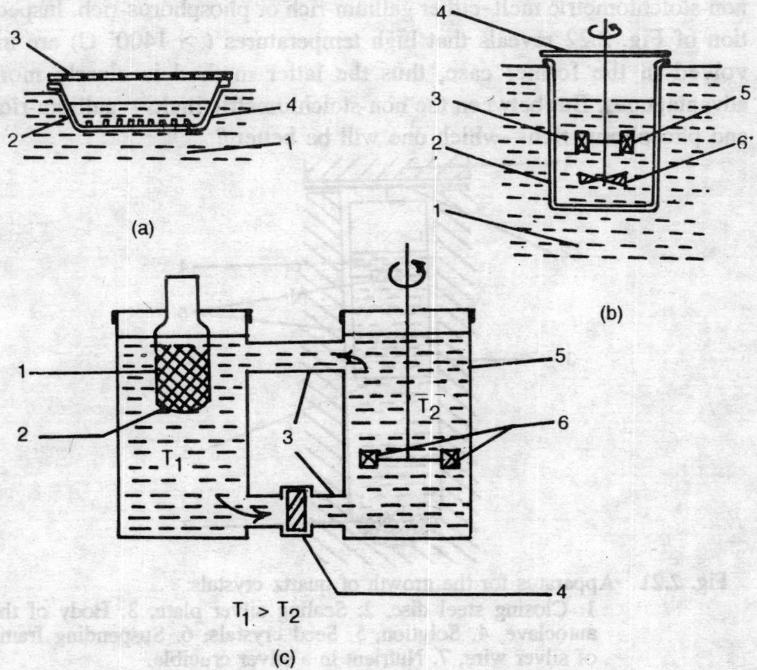

Fig. 2.20. (a) Apparatus for the production of seed crystals:

1. Water bath of the thermostat, 2. Crystallisation vessel, 3. Perfectly or only partially closing lid, 4. Supersaturated solution with the seed crystals.

(b) Growth apparatus:

1. Water bath of the thermostat, 2. Crystallisation vessel, 3. Solution, 4. Perfectly or only partially closing lid, 5. Rotably mounted seed crystals 6. Mixer.

(c) Apparatus for the circulation method:

1. Vessel for the solution supersaturated at temperature T_1, 2. Polycrystalline material in densely woven nylon sack, 3. Connecting tubes, 4. Pump, 5. Growth vessel at temperature T_2, 6. Growing crystals.
(after I. Tarjan and M. Matrai, Eds., *Laboratory Manual on Crystal Growth,* (Akademiai Kiado, Budapest), 1972).

If the temperature is lowered to the point B, nucleation of GaP starts. Since the molar ratio of Ga and P is 1 : 1 in GaP, the melt will get depleted of phosphorus and, hence growth can occur only if the temperature is decreased. A question may be asked whether it is advantageous to grow GaP from its own stoichiometric melt or from a non-stoichiometric melt-either gallium-rich or phosphorus-rich. Inspection of Fig. 2.22 reveals that high temperatures (> 1400° C) are involved in the former case, thus the latter method is clearly more advantageous. But between the non-stoichiometric melts—gallium-rich and phosphorus-rich—which one will be better ?

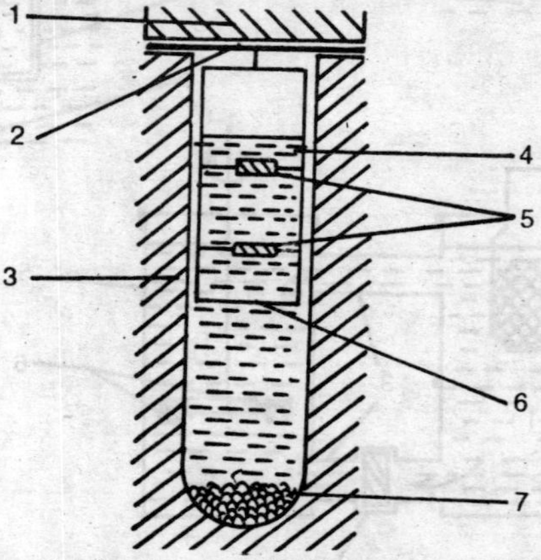

Fig. 2.21 Apparatus for the growth of quartz crystals:
1. Closing steel disc, 2. Sealing silver plate, 3. Body of the autoclave, 4. Solution, 5. Seed crystals, 6. Suspending frame of silver wire, 7. Nutrient in a silver crucible.

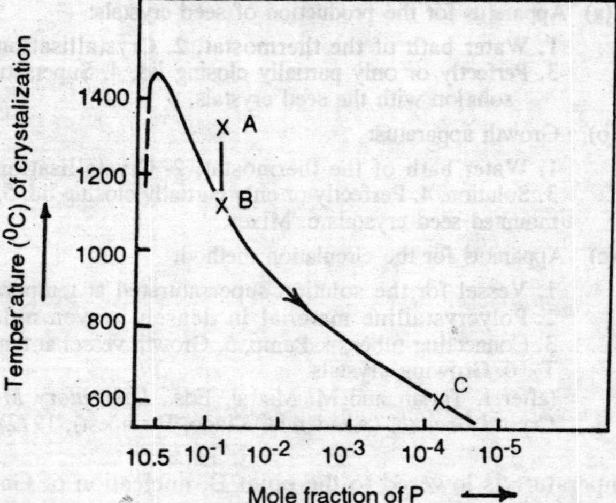

Fig. 2.22. Liquidus curve of the gallium phosphide (GaP) system.

Fig.2.23 gives the pressure-temperature profile of phosphorus in GaP system. The point L corresponds to the maximum in the liquidus curve, i.e., stoichiometric composition. Since phosphorus is more volatile than gallium, the upper branch KL corresponds to the phosphorus-rich melt, while the lower branch ML corresponds to the gallium-rich melt. Thus, a temperature of, say, 1100° C corresponds to two values of equilibrium vapour pressure—100 atm for phosphorus-rich melt and 10^{-3} atm for gallium-rich melt. Since the former requires high pressures, evidently gallium-rich melts are preferable.

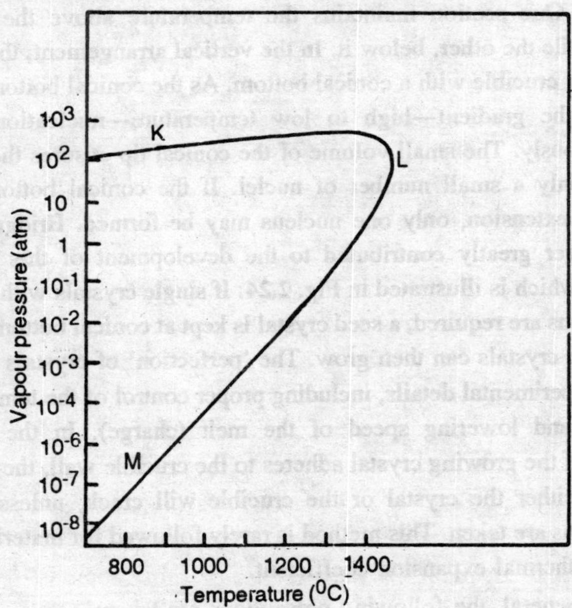

Fig. 2.23. Pressure-temperature profile of phosphorus in the gallium phosphide (GaP) system.

(d) Growth from Melts. This method can be applied to grow crystals which involve a wide range of chemical bonding from ionic, covalent, metallic to molecular, *e.g.*, Zn, Cd, Sn, alkali halides, anthracene, InSb and Ge. An essential requirement in the method is that the material should be stable and not decompose at the melting point. A number of workers, namely, R. Nacken, P.W. Bridgman, D.C. Stockbarger, J. Czochralski, S. Kyropoulos and A. Verneuil, have contributed to the development of this method. Its variants include normal freezing, zone melting and crystal pulling.

(i) ***Normal Freezing.*** This method, consists of unidirectional freezing. The freezing is effected in a temperature gradient zone, in which the temperature changes gradually from slightly above the melting point (of the charge) to slightly below it. The charge is exposed to the gradient zone either by moving it through the gradient or alternately, the gradient is moved. The movement of the charge may be horizontal or vertical. In order to control the temperature profile, the heating furnace is divided into two sections with separate heaters and controls. One section maintains the temperature above the melting point, while the other, below it. In the vertical arrangement, the charge is put in a crucible with a conical bottom. As the conical bottom moves through the gradient—high to low temperature—nucleation occurs spontaneously. The small volume of the conical tip ensures the formation of only a small number of nuclei. If the conical bottom has a capiliary extension, only one nucleus may be formed. Bridgman and Stockbarger greatly contributed to the development of this gradient method, which is illustrated in Fig. 2.24. If single crystals with specific orientations are required, a seed crystal is kept at conical bottom around which the crystals can then grow. The 'perfection' of crystals depends on the experimental details, including proper control of the temperature gradient and lowering speed of the melt (charge). In the vertical method, if the growing crystal adheres to the crucible wall, the chances are that either the crystal or the crucible will crack, unless special precautions are taken. This method is rarely followed for materials with negative thermal expansion coefficient.

In general, the following precautions are taken:

(a) To avoid contamination of the melt, all the materials coming in contact with the melt or its vapour, must be carefully selected.

(b) The temperature of the melt should always be only slightly above the melting point, in order to avoid melt decomposition and its possible reaction with the crucible.

(c) Nucleation should occur under large temperature gradient conditions.

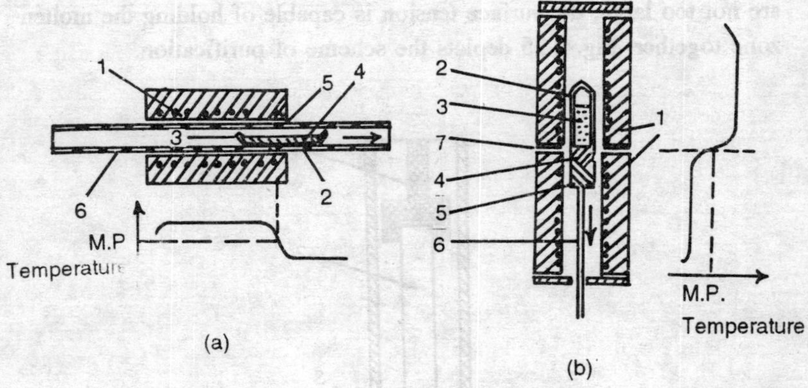

Fig. 2.24. (a) Schematic diagram of the horizontal freezing method; below: ideal temperature distribution.
1. Resistance furnace, 2. Boat. 3. Melt, 4. Crystal, 5. Isotherm belonging to the melting point, 6. Heat insulating tube

(b) Schematic diagram of the vertical normal freezing method; right: ideal temperature distribution.
1. Resistance furnace assembled from two parts, 2. Crucible, 3. Melt, 4. Crystal, 5. Crucible support, 6. Lowering rod, 7. Baffle.

(d) Growth should occur under small temperature gradient conditions in order to avoid mechanical stresses.

(e) The growth rate should be low but not too low. Small growth rates can produce strain-free, good quality crystals, but too small growth rates are uneconomical and may also increase the chances of melt contamination.

(f) The entire volume of the grown crystal should be maintained in a uniform temperature zone and cooled slowly to room temperature.

(ii) ***Floating Zone Method.*** This is a variant of the zone melting/refining technique, and can be used for the purification of semiconductors like silicon and germanium. It is especially suitable for situations in which we wish to avoid contamination of the melt by the crucible material.

A sintered rod, clamped at both ends, is held vertically in vaccum or inert gas; a short section of the material is melted, mainly through induction heating, as in zone refining. If the zone length and diameter are not too large, the surface tension is capable of holding the molten zone together. Fig. 2.25 depicts the scheme of purification.

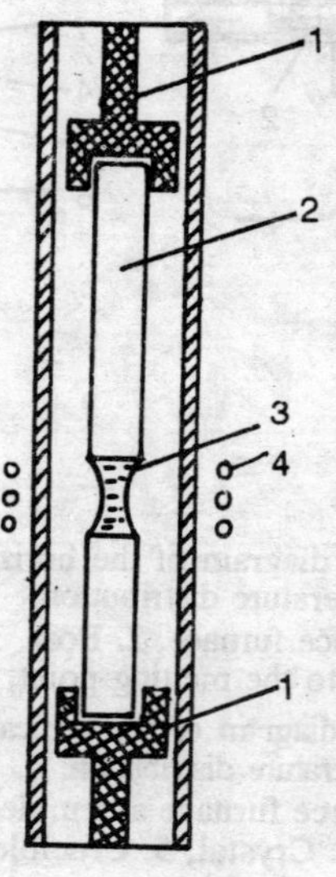

Fig. 2.25. Schematic diagram of an apparatus for refining silicon by the method of 'floating zone'.
1. Terminals, 2. Silicon specimen, 3. Molten zone, 4. Heater (inductor) (after V.A. Bruk et al, in *Semiconductor Technology* (Mir Publishers, Moscow).

(iii) ***Crystal Pulling.*** In this method, the charge is melted and maintained at a temperature slightly above the melting point; the pulling rod is lowered to just touch the melt. Since the rod is at a lower temperature, preferential condensation of the melt occurs at the pointed tip of the pulling rod. The crystal is then 'pulled' slowly. The rate of pulling depends upon various factors such as thermal conductivity, latent heat of fusion of the charge and the rate of cooling of the pulling rod. Often 'seed' crystals are attached to the tip of the pulling rod to grow crystals with specific orientations. Crystal pulling is an art and many important materials such as silicon, germanium and indium antimonide (InSb) can be grown by this method. Nacken, Czochralski and Kyropoulos have contributed to the different variants of this method. Fig. 2.26 shows the essentials of this technique.

Sometimes heating of the charge is done by the electric arcs in the so-called 'cold crucible' technique. This method was popularised by J.C.C. Fan and T.B. Reed. The main advantage of this method is that the thin layer of the unmolten charge in contact with the water-cooled hearth prevents contamination from the crucible material.

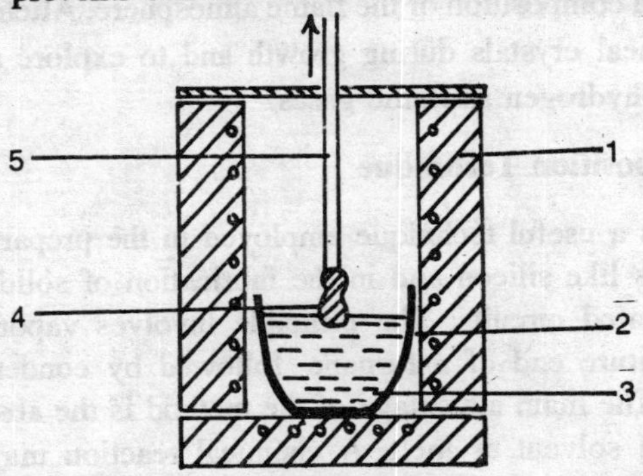

Fig. 2.26. Apparatus for the pulling method:
1. Furnace (heating of sides and bottom can be regulated separately), 2. Crucible, 3. Melt, 4. Seed crystal, 5. Cooling tube.

Materials with high melting points (upto 2000° C) such as ThO_2 , Al_2O_3 and ZrO_2 can be conveniently grown by this technique. Honig and coworkers have highlighted the utilization of arc techniques in the synthesis of inorganic materials.

(e) Flame Fusion Method. This method was developed by Verneuil. It is mainly used for making synthetic gems like ruby, sapphire, spinel, etc., and for watch/instrument bearings and knife edges for balances. The charge is taken in a finely powdered form and carried

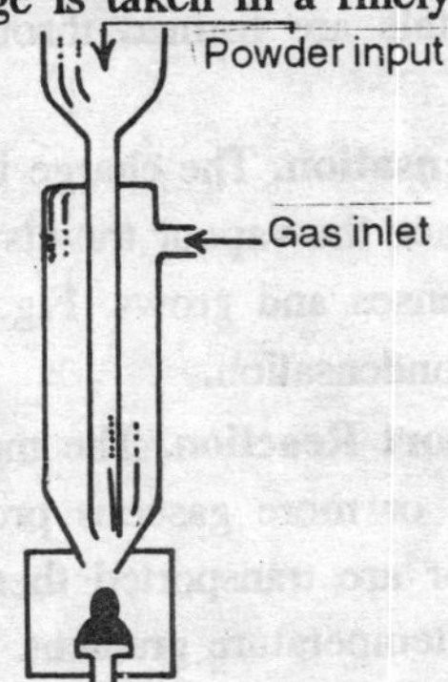

Fig. 2.27. Crystal growth from the melt by the 'Verneuil' method.

to a burner by oxygen. There it is melted in an oxy-hydrogen flame and a single crystal grows around a seed crystal. Fig. 2.27 shows the apparatus generally employed.

Since no crucible is used, crystals of high-melting materials, for which an appropriate non-reactive crucible material has not been found, can be grown. The chief disadvantages of the method are high thermal gradients and composition of the flame atmosphere. Attempts have been made to anneal crystals during growth and to explore alternatives to oxygen and hydrogen as flame gases.

Vapour Deposition Technique

This is a useful technique employed in the preparation of high-purity metals like silicon and in the fabrication of solid state devices and inter grated circuits. The principle involves vaporisation at the high-temperature end of a furnane, followed by condensation at the colder end. The main advantage of the method is the absence of interference from solvent or melt. A chemical reaction may or may not occur when the gaseous species are travelling from the hot to the cold end. The vaporisation may be achieved through either evaporation, sputtering, plasma or chemical reactions. When the layer deposition is done on an already existing substrate surface, it is called epitaxial growth. Silicon is produced by the thermal decomposition of silane (SiH_4) or reduction of chlorosilanes with hydrogen, *e.g.*,

$$SiH_4(g) \xrightarrow{\text{carrier gas}} Si(s) + 2H_2(g)$$

$$SiCl_4(g) + 2H_2(g) \rightarrow Si(s) + 4HCl(g)$$

In nature, snow crystals are formed through condensation of vapour.

(f) Growth by Condensation. The charge is sublimed at a high temperature end of the furnace the vapour travels to the colder end of the furnace, where it condenses and grows. Fig. 2.28 (a) shows the simple case of growth by condensation.

(g) Chemical Transport Reaction. The material to be crystallized is converted into one or more gaseous products, which either diffuse to the colder end or are transported there by a transporting (carrier) gas by means of a temperature gradient. At the cold end, the

Table 2.1. Modern Applications of Artificially Grown Single Crystals

No.	*Single crystal*	*Applications*
1.	Si, Ge, GaAs	Transistors
2.	GaAs	Tunnel diodes, parametric diodes, signal diodes
3.	Si, Ge (As, P)	Strain gauges
4.	YIG	Microwave limiters and tunable filters
5.	$CaWO_4$, CaF_2 ruby, GaAs, InP, InSb, InAs, Ge (As, P)	Masers and lasters
6.	Quartz, Rochelie salt, CdS, GaAs, ADP	Electromechanical transducers
7.	Quartz	Filters and oscillators
8.	CaF_2, LiF, quartz, calcite	Optical uses
9.	Anthracene, KCl, Si, GaAs, NaI(Tl), Ge(Li), TGS	Radiation detectors
10.	CdS	Ultrasonic amplifiers
11.	Sapphire	Industrial bearings
12.	Diamond, SiC, sapphire	Cutters and abrasives
13.	Si and Ge	Rectifiers
14.	GaP, GaAs, Ge(As, P)	Electroluminescent devices
15.	KDP, $LiNbO_3$, $LiTaO_3$ barium sodium niobate, barium strontium niobate	Laser modulators, harmonic generators and parametric devices
16.	$BaTiO_3$	Ferroelectrics

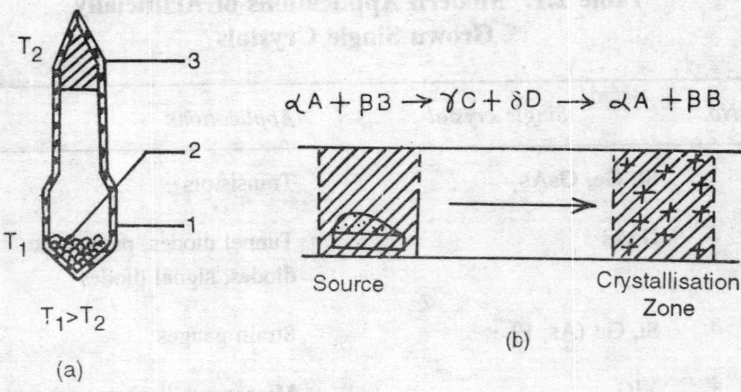

Fig. 2.28. (a) Apparatus for the simple case of growth by condensation: 1. Closed glass or quartz ampoule, 2. Polycrystalline material, 3. Single crystal.

(b) Apparatus for the growth by chemical transport reaction.

reaction is reversed so that the gaseous product(s) decompose to deposit the parent material, liberating the transporting agent, which diffuses to the hotter end and again reacts with charge. One of the hazards of this method is the enormous pressure likely to be built up in a closed system, which may result in an explosion. Fig. 2.28 (b) schematically depicts the apparatus used for the growth of crystals by chemical transport reaction.

Table 2.1 lists some of the applications of artificially grown single crystals.

3

The Law of the Constancy of Angles, and the Measuring of Crystals

3.1. Introduction

We visualize a crystal as a solid substance of reticular structure (Fig. 3.1). The particles of which it is made up constitute a crystal lattice, where faces correspond to planes, edges to rows of nodes, and coigns to nodes.

This should not be understood literally. If a microscope could be designed to show us directly the atoms and molecules, we should see that the structure of a real crystal is very much unlike that of a geometrically correct *space* lattice.

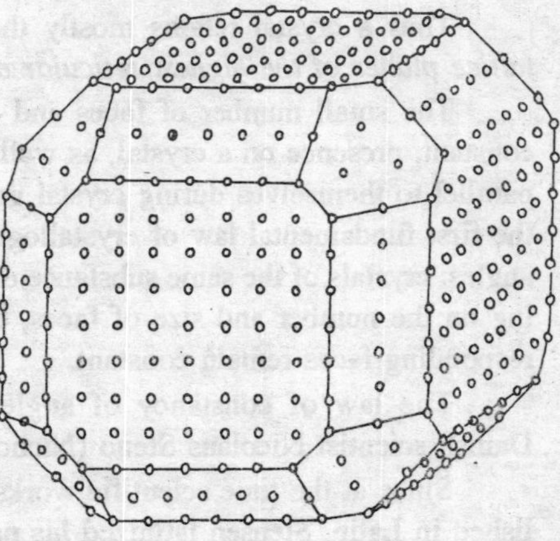

Fig. 3.1. Structure of crystal.

We shall discuss this latter, in more detail, but for the present we shall adopt as a general scheme a lattice affording easy understanding and explanation of specific properties of crystalline substance.

Insofar as three nodes from different rows determine the position of each plane of a lattice, or of a plane net, as it is conventionally called, it is evident that every lattice must have an infinite number of possible planes definitely positioned in space. Some planes are thickly covered with nodes, others not so thickly. Therefore, we can distinguish plane nets by *their degree of density, i.e., by the number of nodes per unit surface area (reticular density).*

If all the possible planes of a crystal lattice (their number being infinitely large) found reflection in the faceting of a crystal it would be difficult to find simple laws governing the faceting. On the other hand, it might seem that several crystals of the same chemical compound, *i.e.*, crystals built according to the same lattice, might have entirely different habits and would have no two identical interfacial angles.

Actually the picture is quite different.

The crystals do not develop all the possible faces, but just a few of them. What faces are these and why ? The following *low of consecutive development of crystal faces* can be formulated: habit faces, *i.e.*, those appearing most frequently, are developed *in the order of decreasing densities of their nets.* Thus, the first faces appearing on a crystal have the maximum density, the ones appearing later have progressively decreasing densities. This can be roughly explained by the higher mechanical strength of faces with maximum densities and by their slower growth. Different substances display some very peculiar features in this respect. For instance, NaCl crystals have very few different faces (only 3 or 4 types); while other minerals, such as calcite, are richly faceted, the faces having nets of highly diverse densities.

Thus a crystal retains mostly those faces *which correspond to lattice planes of the highest reticular density.*

The small number of faces and their frequent, sometimes even constant, presence on a crystal, as well as their ability to be translated parallel to themselves during crystal growth, has served as a basis for the first fundamental law of crystallography—the law of constancy of angles: crystals of the same substance can have different habits depending on the number and size of faces, but the angles between the corresponding faces remain constant.

The law of constancy of angles was first formulated by the Danish scientist Nicolaus Steno (Nicholas Stensen) in 1969.

Since at the time scientific works were usually written and published in Latin, Stensen latinized his name. His work was called "De solido inter solidum naturaliter contento", *i.e.*, "On a solid naturally contained in a solid".

The first edition was printed in Florence in 1669, the second ten years later in Leiden.

The constancy of interfacial angles was demonstrated by Stensen for two substances: quartz (SiO_2) and hematite (Fe_2O_3).

The validity of the law of constancy of angles for crystals of all substances was confirmed by the French scientist Rome Delisle much later, in 1783.

It should only be added that the constancy of angles takes place under identical temperature and pressure conditions.

It can be seen from the above that the angles of crystals, or the mutual inclination of crystal faces, are highly important. It is natural, therefore, that one of the first problems in the study of the external form of a crystal is the measurement of its interfacial angles.

This is accomplished with the aid of devices called *goniometers*. Different types of goniometers are used depending on the size of the crystal, as well as on the number and quality of its faces. The most widely used instruments are:

1. Contact Goniometer. It consists of a protractor and a rotating bar pivoted about the centre of the protractor (Fig. 3.2). In the measurement of an interfacial angle the goniometer is applied to the crystal so that the edge between faces is parallel to the axis of rotation of the bar. Such instruments are accurate to within 30 min., and are used when a more accurate measurement is impossible, usually because of the size of the crystal or of the imperfection of its faces.

The first contact goniometer was built by Carrangeot in 1772. At the present time measurements of large crystals are made with improved instruments of this type.

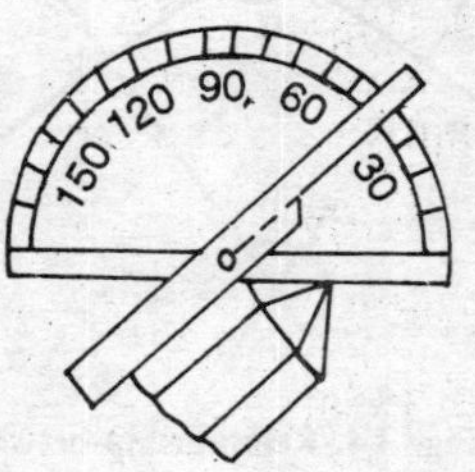

Fig. 3.2. Contact goniometer.

2. One-circle Reflecting Goniometer is designed on the principle of obtaining consecutive reflections of rays of light from crystal faces. It is shown diagrammatically in Fig. 3.3. A crystal K is mounted with the aid of waa on a holder connected to a stage which rotates together with a graduated circle *H*. The edge between the faces being measured is placed parallel to the axis of rotation. Rays from a source of light pass through a collimator* C and strike the

* A collimator is that part of an optical instrument which produces a beam of parallel rays.

crystal in a parallel beam. The observer looking through a telescope *F* sees the image of the source of light, called a signal, only when one of the crystal faces (a_1) occupies the position shown in Fig. 3.3, *i.e.*, is normal to the bisectrix of the angle between the optic axes of the collimator and the telescope.

Having taken a reading with the aid of a nonius, corresponding to the required position, and having rotated the crystal together with the graduated circle so that the face a_2 takes the place of the face a_1 (or lies parallel to it), we can find the angle between normals N_1 and N_2 by the difference in readings. This angle, β (Fig. 3.4), is complementary (to 180°) to angle α between the faces a_1 and a_2. A complete revolution of the crystal makes it possible to measure all the angles of one zone.

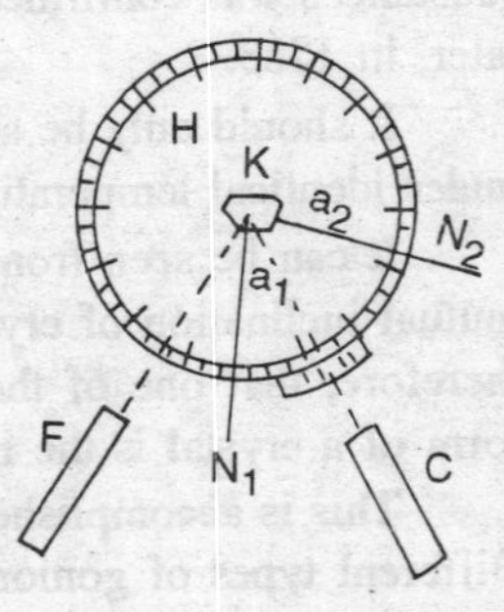

Fig. 3.3. Function of the reflecting goniometer.

The one-circle goniometer allows measurements to be made with an accuracy of 1. or higher. The first instrument of this type, invented in 1809 by W.H. Wellaston, had a vertical graduated circle. It had neither a telescope nor a collimator, and afforded a much lower accuracy of measurement. The principal drawback of the one-circle reflecting goniometer is the impossibility of measuring interfacial angles of more than one zone at one setting of a crystal. When a crystal 3-4 mm in size has a large number of faces (25-30), measurement by zones is complicated and labour-consuming. The crystal has to be reset several times, this involving repeated mounting. In each mounting the crystal has to be set up so that some faces from one or two previously measured zones might be included. This increases the difficulties in mounting. To avoid repeated mounting, a *two-circle reflecting goniometer,* or a *theodolitic goniometer,* was designed. It has two mutually perpendicular axes of rotation (Fig. 3.5) and accordingly two graduated circles, a telescope, and a collimator. The position of a crystal corresponding to a signal in the telescope can be noted by taking readings off the two graduated

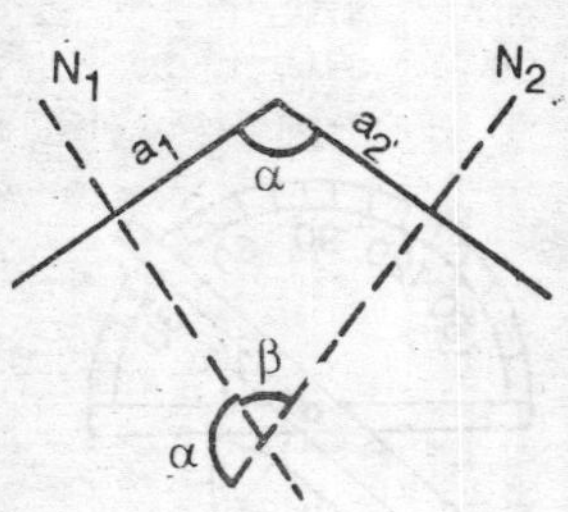

Fig. 3.4. Relationship between the interfacial angle and the angles between normals

circles. Thus, two spherical coordinates are obtained for each face, this allowing to determine its position in space.

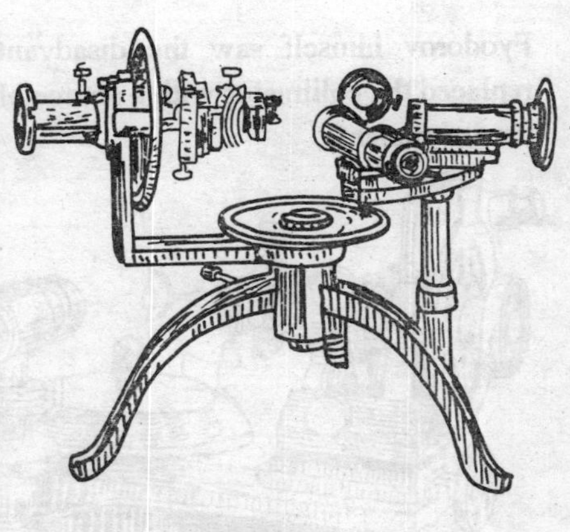

Fig. 3.5. Two-circle goniometer.

The idea of a two-circle goniometer was first conceived by professor W. H. Miller (1801-1880) of Cambridge University. In 1874 he had to measure a platinum bead which was obtained in the flame of a blowpipe. The globular bead, only 0.73 mm in diameter, had 147 faces. Professor Miller combined a one-circle goniometer of the Babinet type with a small Wollaston goniometer so as to obtain mutually perpendicular axes of rotation. This goniometer is still at the mineralogical museum of the Cambridge University. 15 years later, in 1989, a genuine two-circle goniometer was built by professor Y.S. Fyodorov.

The Fyodorov two-circle goniometer (Fig. 3.6) had no collimator, and an autocollimation device was installed in the telescope. But soon

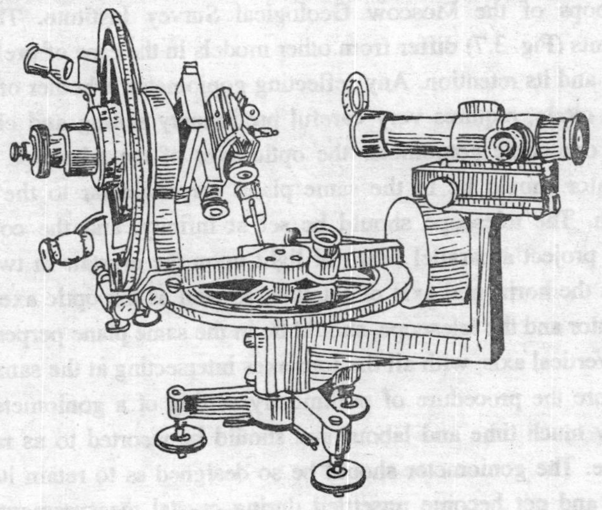

Fig. 3.16. Fyodorov goniometer.

Fyodorov himself saw the disadvantages of this arrangement and replaced the collimator with a source of light sufficiently removed from the instrument.

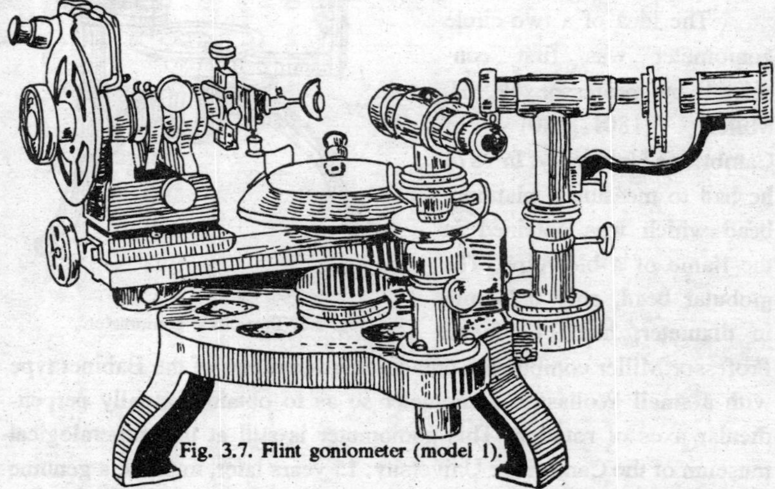

Fig. 3.7. Flint goniometer (model 1).

Fyodorov's work on the design of two-circle goniometers was carried on by Y.Y. Flint. In 1938-1939 he succeeded in organizing the manufacture of two-circle goniometers of his own design in the workshops of the Moscow Geological Survey Institute. These instruments (Fig. 3.7) differ from other models in the ease of preliminary setting and its retention. Any reflecting goniometer, whether one-circle or two-circle, requires very careful preliminary setting and checking. In the one-circle goniometer the optic axes of the telescope and the collimator should lie in the same plane perpendicular to the axis of rotation. The telescope should be set at infinity, and the collimator should project a parallel beam of light upon the crystal. In two-circle models the horizontal axis of rotation, as well as the optic axes of the collimator and the telescope, should lie in the same plane perpendicular to the vertical axis, with all the four axes intersecting at the same point. Therefore the procedure of preliminary setting of a goniometer takes up very much time and labour and should be resorted to as rarely as possible. The goniometer should be so designed as to retain its initial setting and not become unsettled during crystal measurements. Y.Y. Flint's models are satisfactory in this respect.

Besides the instruments of Russian design, there are two-circle goniometers of Czapski and of Goldschmidt. Goldschmidt's instruments have a very well-designed and convenient telescope.

The use of two-circle goniometers considerably simplified the procedure of crystal measurement. They were first introduced in 1893 and were soon accepted universally.

Besides the above-mentioned goniometers designed for measurements of medium-sized crystals (1 to 5-6 mm), there are instruments for measuring very large or very small crystals. There are also goniometers for special purposes, *e.g.*, for measuring crystals at very high and very low temperatures, for making thin oriented sections, for measurements of crystals in a solution during growth.

4

Metallic Bonding, Semiconductors and Superconductors

4.1. Metallic Bonding

Introduction. The geometric arrangement of atoms in a metal crystal is the most compact one. *The forces that keep the atoms so closely bound together in a metallic crystal constitute what is generally known as the metallic bond.* The various characteristics of the metallic bond are:

(i) The metallic bond is essentially a covalent bond, without saturation, allowing a large number of atoms to be held together by a mutual sharing of valency electrons.

(ii) The density of electrons between the atoms is much lower than is allowed by the Pauli's exclusion principle. This allows electrons to move freely from point to point without a significant in crease in their energy.

However, the metallic bond differs from the covalent bond in the following respects:

(i) The valency electrons in a covalent bond are *localized* and, therefore, a covalent bond has a *directional character* while in a metallic bond the valency electrons are spread all over the crystal more or less uniformly. The metallic bond therefore is *non-directional* in character.

(ii) The valency electrons in a covalent bond are attracted strongly towards the nuclei of the atoms while the valency electrons in a metallic bond, being mobile, do not experience such strong forces of attraction towards the nuclei. *Metallic bonds are, therefore, weaker*

than the covalent bonds. Due to this reason, the energy required to convert a mole of a metal (say, copper) into free atoms is smaller than the energy required to vaporise a mole of a covalent substance into free atoms (say, graphite).

1. Classical Free Electron Theory or Electron Sea Theory

This theory was put forward by Drude and later developed by Lorentz. According to this theory,

(i) All the atoms in metal have several unoccupied electron orbitals in their outer shells.

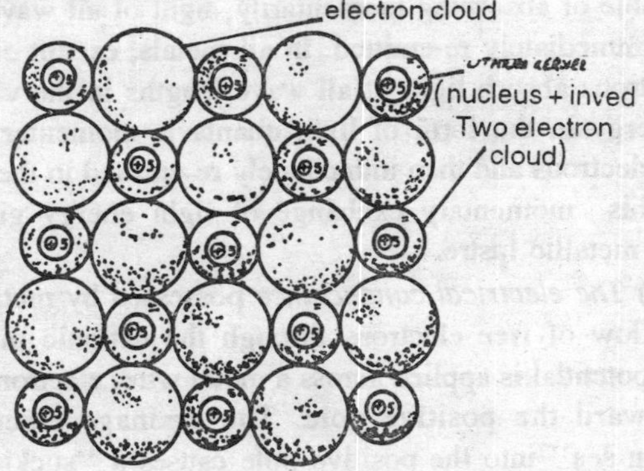

Fig. 4.1

(ii) Due to the low ionisation energies of metals they have a tendency to losse some of their valency electrons and form positive ions called *metal nuclei.*

(iii) These electrons are not attached to any particular metal ion but are free to move like the gas molecules throughout the volume of the metal without any restriction. The availability of a large number of vacant orbitals in the metal atoms helps the electrons in this free movement. Thus the electrons are not bound to a single nucleus but spread out around several nuclei in the form of a charge cloud. Such electrons are said to be *delocalised.* Thus, a metal can be compared to an assembly to cations, immersed in a sea of electrons. For this reason, this model is also referred to a *electron sea model.*

(iv) Due to the close packing of atoms in the metal, it is assumed that several electron charge clouds are attracted by several nuclei. As a result of this attraction, each nucleus comes nearer to several electron charge clouds to produce low potential energy. This leads to big force of attraction between the nuclei of metal atoms and surrounding

electrons. The strong attraction of several nuclei for several electron clouds gives a solid stable structure of metal. For example, Fig. 4.1 **represents electrons sea model for lithium metal.**

Successes of Free Electron Theory. Though the Drude-Lorentz theory gives a greatly simplified picture, it does nevertheless, afford a useful explanation for many of the properties of metals which are given below:

(*a*) *The optical properties* of metals, such as metallic lustre and silver white or silver-grey colour are due to the fact that free electrons are capable of absorbing momentarily, light of all wavelengths, which is then immediately re-emitted. In all metals, except copper and gold, the electrons absorb light of all wave-lengths in the visible spectrum. This energy in the form of light quanta is momentarily accepted by mobile electrons and then immediately re-emitted in the form of visible light. This momentary exchange of light energy gives rise to the familiar metallic lustre.

(*b*) *The electrical conductance* possessed by metals is obviously due to flow of free electrons through the cationic matrix. When an electric potential is applied across a metal wire, electrons from the wire flow toward the positive pole. The drainage electrons from the "electron sea" into the positive pole causes a "sucking in" of fresh electrons from the negative pole. Thus, the passage of the electric current, which means that the flow of electrons from negative to the positive pole, continues.

The conductivity of metals decreases with rise in temperature. This is because as the temperature of a metal sample is raised, the positively charged kernels move rapidly because of the rise in their kinetic energy. This movement of the kernels will interfere with the flow of electrons through the metal sample. Thus as a result, the resistance of the metal to the flow of charge would increase as the temperature of the metal rises or in other words, the conductivity would decrease.

(*c*) *Thermal conductivity of metals. Materials which are good conductors of electric current are also good conductors of heat.*

Heat transferred through a metallic piece takes place whenever one side of it is at higher temperature as compared to the other side. Higher temperature on one side of the metal pieces implies that kernels in that sides possess a higher kinetic energy as compared to the kernels on the other side. Rise in kinetic energy of kernels changes the kinetic energy of delocalised electron because the vibrating kernels transfer kinetic energy to kernels on the far away parts of the crystals (Fig. 4.2) (*d*) *The mechanical properties of metals* (such as melleability, ductility and ability to bend or get deformed without breaking) are consistent with the concept of metallic lattice. The deformation of the crystal lattice due to imposed stress get relieved by the slippage of crystal planes as shown in Fig. 4.3. One plane of atoms slips over another, and the mobility of electrons enables them to immediately re-establish a bonding between the two planes. This sort of thing is

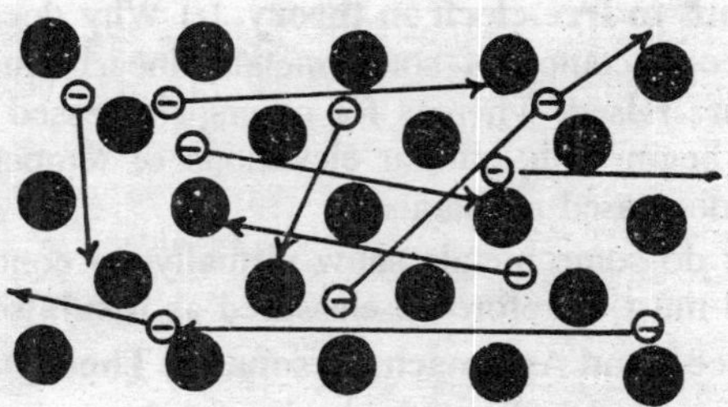

Fig. 4.2

not possible in non-metals, where the repulsive forces of like charges would cause a shattering of the crystal lattice (Fig. 4.4)

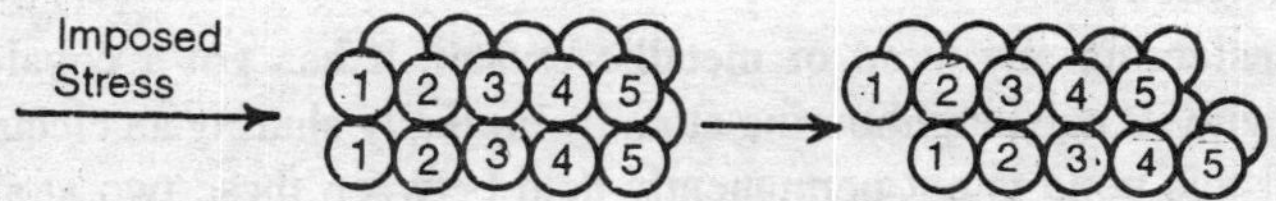

Fig. 4.3.

(*e*) *High tensile strength.* The metals have high tensile strength as they resist stretching without breaking. This can be explained on the basis of strong electrostatic attraction between mobile electrons and

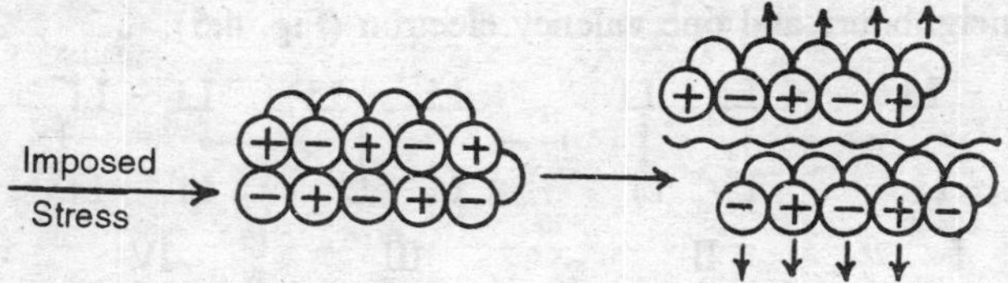

Fig. 4.4

the positively charged metal ions immersed in them. The covalent substances do not have high tensile strength because the electrostatic force of attraction is not there as there are no oppositely charged units in the latter.

(*f*) *Elasticity.* In case of certain substances when force is applied, the substance gets deformed and when this deforming force is removed, the substance regains its original state. This property possessed by certain substances is known as elasticity. Metals are highly elastic. This is because in case of metals when deforming force is applied, layers of atoms slip over each other and when the deforming force is removed, the atoms go back to their original state almost completely.

Objections to free electron theory. (*i*) Why does the resistance to electrical conductance of some metals linearly increase as their temperatures are raised, whereas for others, possessed relatively high resistances to begin with, similar elevations of temperature result in exponentially decreased resistance ?

(*ii*) Why do some metals show virtually no conductance in the solid state and must therefore be classified as **insulators** ?

2. Valence Bond Approach (Resonance Theory). An alternative view of the metallic bond, on the basis of valence bond theory, was developed principally by Pauling. It treats metallic bonding as being essentially covalent in origin and pictures the metallic structure as involving resonance of electron-pair bonds between each atom and its nearest neighbours.

Considering any atom of metallic crystal, it has got a covalent bonding with its one neighbouring atom by mutually sharing an electron each. But this bond is not permanently held between these two atoms, instead it is continuously in resonance. At one moment it was in between above said two atoms, at second moment one of the atom is already shared by other neighbouring atom. At the third moment, same atom is having its bonding with any third neighbouring atom. Let us take the example of lithium in which each lithium atom has eight nearest neighbours and one valency electron (Fig. 4.5).

```
Li - Li          Li    Li            Li⁺    Li          Li - Li⁻
          ←→     |     |      ←→            |     ←→           |
Li - Li          Li    Li            Li - Li⁻           Li⁺   Li
   I                II                   III                 IV

                 Li    Li⁺           Li⁻ - Li
          ←→     |            ←→     |
                 Li⁻   Li            Li    Li⁺
                    V                   VI
```

Fig. 4.5

Though only four atoms are shown in these pictures, the actual bonding includes all of the atoms of the crystal and is three dimensional.

Properties explained by valence bond approach. In the first long period, the cohesive energies, melting points, boiling points, heats of fusion and hardness increase whereas interatomic distances decrease from potassium to chromium. These properties remain constant from chromium to nickel and with the exception of interatomic distance, they steadily decrease from copper to barium. The similar behaviour has also been observed in second and third long periods if lanthanides and actinides are also considered.

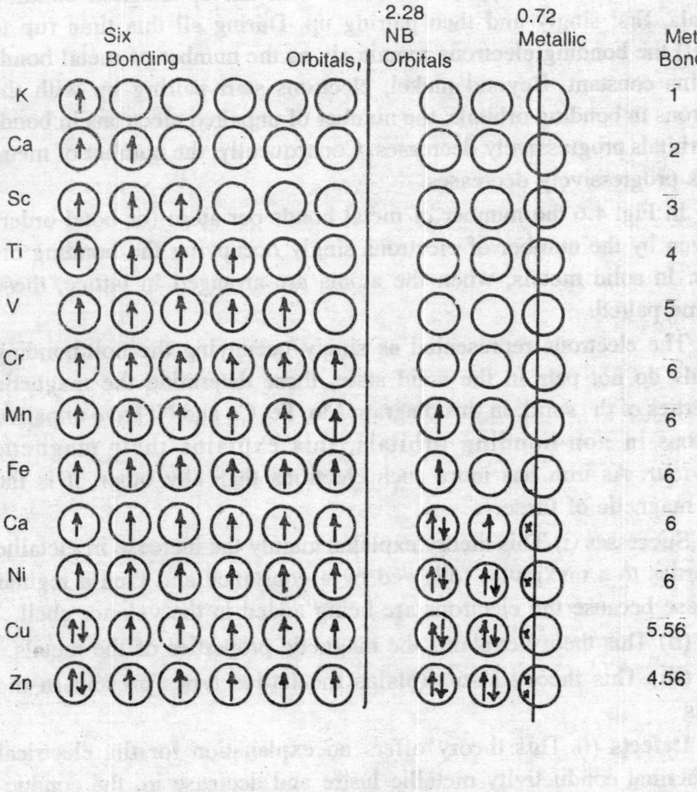

	Six Bonding Orbitals,	2.28 NB Orbitals	0.72 Metallic	Metal Bonds
K				1
Ca				2
Sc				3
Ti				4
V				5
Cr				6
Mn				6
Fe				6
Ca				6
Ni				6
Cu				5.56
Zn				4.56

Note : ↑ and × indicate partial filling of orbitals.

Fig. 4.6. The orbitals filling and number of metallic bonds per atom for K through Zn.

In order to explain the above mentioned properties, the orbitals involved in bonding are not only five *(n—1) d* and one *ns* but three *np* as well, or nine orbitals per atom are made available for bonding. Out of these six are the *bonding orbitals,* 2.28 the *non-bonding orbitals* and 0.72, *the metallic orbitals.* The empirical value of 0.72 indicates that at a given instant not all atoms have equivalent configuration because of resonance. For example, if one hundred atoms are available, the total number of orbitals are 900. Out of these, six hundred orbitals being used as bonding orbitals, 228 are non-bonding and 72 metallic bonding at any instant. The orbital filling and the number of metallic bonds per atom are sown in Fig. 4.6 from potassium to zinc.

In Fig. 4.6 while filling, after six electrons have entered six bonding orbitals (chromium), the electrons then fill up the non-bonding orbitals, first singly and then pairing up. During all this time (up to nickel) the bonding electrons remain six as the number of metal bonds remains constant. Beyond nickel, electrons start pairing up with the electrons in bonding orbitals, the number of unpaired electrons in bonding orbitals progressively decreases. Consequently, the number of metal bonds progressively decreases.

In Fig. 4.6 the number of metal bonds per atom (or bond order) is given by the number of electrons singly occupying the bonding orbitals. In solid metals, when the atoms are arranged in lattice, these become paired.

The electrons represented as singly occupying the non-bonding orbitals do not pair in the solid state; these determine the magnetic properties of the solid. In this diagram Mn, Fe, Co and Ni have unpaired electrons in non-bonding orbitals, this explains their magnetic behaviour. As iron has more such electrons than any other, it is the most magnetic of these.

Successes (i) This theory explains mainly the increase in metallic properties to a maximum followed by a constancy and then a regular decrease because the electrons are being added in the valence shell.

(ii) This theory explains the magnetic properties of the metals.

(iii) This theory also explains the lattice types present in the metals

Defects *(i)* This theory offers no explanation for the electrical and thermal conductivity metallic lustre and decrease in, the conductivity with rising temperature.

(ii) This theory offers no explanation for the semiconduction.

3. Molecular Orbitals Approach (Band Theory). This theory takes into account that all the electrons off all atoms in the metallic crystal belong to the crystal as a whole and the metallic bonding results from the delocalisation of the free electron orbitals over all the atoms of a metal structure.

When two atomic orbitals combine, two molecular orbitals are formed. One of these is of lower energy and is called *bonding molecular orbital* while the other is of higher energy and is called *antibonding molecular orbital.* This can be illustrated by the formation of Li_2 molecule from two Li atoms. Ignoring any inner electrons, the 2s atomic orbitals in each of the two Li atoms combine to form two molecular orbitals, *i.e*, one bonding and other antibonding. The valency electrons occupy the bonding molecular orarbital Fig. 4.7 (*a*).

Let us now consider the formation of Li_3 from three Li atoms. Three 2*s* atomic orbitals would combine to form three molecular orbitals, *i.e.*, one bonding, one non-bonding and one antibonding. The energy of the non-bonding molecular orbital lies between that for the bonding and antibonding orbitals. The three valency electrons of the

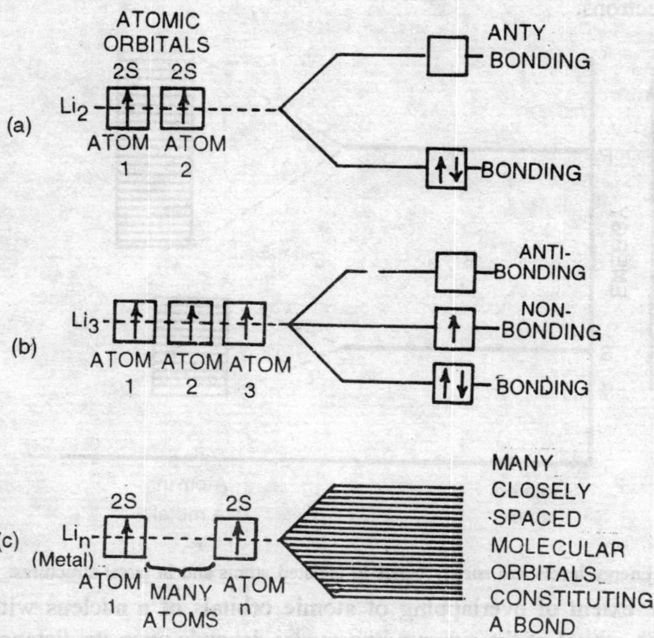

Fig. 4.7. Development of molecular orbitals into bonds in metals.

three Li atoms would occupy the bonding molecular orbital (2 electrons) and the non-bonding molecular orbital (1 electron) [Fig. 4.7(*b*)].

With the increase in the number of atoms in the cluster, the spacing between the energy levels of the various orbitals decreases further and with a large number of atoms the energy levels of the orbitals are close together that they virtually form a continuum Fig. 4.7 (*c*).

In general, *n* atomic orbitals may combine to form *n*/2 bonding molecular orbitals and *n*/2 antibonding molecular orbitals. In a metallic crystal, there is a very large number of atoms of the order of 10^{20}. When these atoms (10^{20}) combine, 10^{20} molecular orbitals will be formed. As this is very large number the energy separation between the various molecular orbitals will be very small and this yields the formation of energy bands. These bands may be separated from each other as in Fig. 4.8. In such a case the gaps between the bands represent energies in which the electrons connot be present. Such gaps are therefore called *forbidden bonds*.

Some energy bands in a solid may overlap, as in Fig. 4.9 and in such a case there is then a continuous distribution on allowable energies for its electrons.

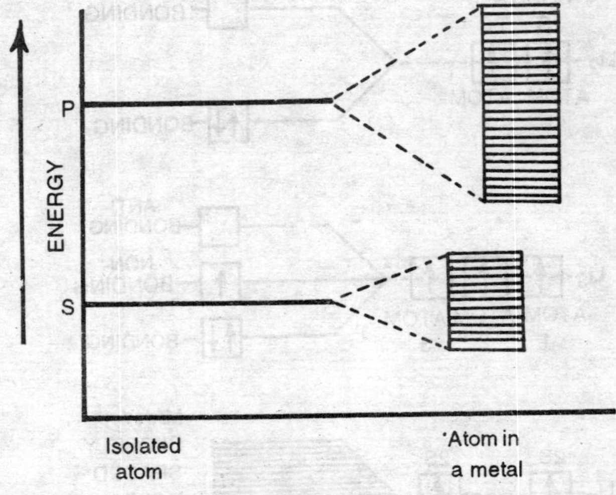

Fig. 4.8. Energy levels and energy bands in isolated atoms and in metal structures.

The extent of overlapping of atomic orbitals of a nucleus with the atomic orbital of the surrounding nuclei depends upon its distance from these surrounding nuclei, *i.e.*, the internuclear distance. If the internuclear distance increases, there is lesser overlapping and, thus there is less of perturbation and orbitals remain essentially atomic. As the internuclear distance decreases, the amount of overlapping of

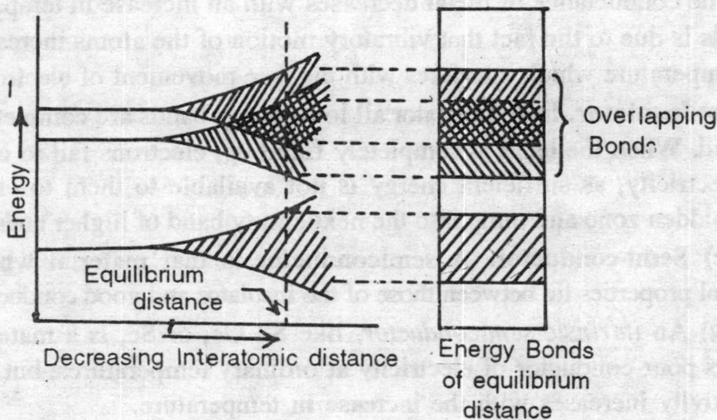

Fig. 4.9

atomic orbitals with atomic orbitals of neighbouring nuclei increases. Thus, there is more of perturbation and overlapping orbitals become molecular orbitals. The filling of these bands and the width of forbidden zones determine whether a substance is a conductor, insulator, or a semi-conductor. Let us discuss these one by one.

(a) True metal or conductor. The true metal is characterised by excellent conductance. If shows increasing resistance as temperature rises. The distinctive feature of a true metal (conductor) is either a *partially filled band* or *an unoccupied band overlapping an empty band.*

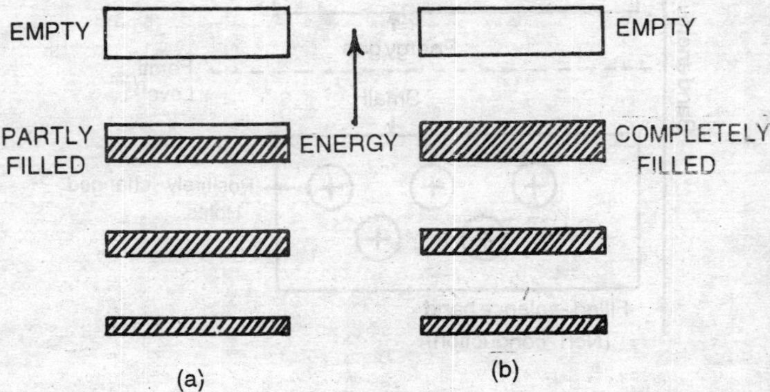

Fig. 4.10. Electrons occupation in the bands in (*a*) a conductor, (*b*) an insulator.

On the application of potential difference to a metal, the electrons in a partially filled energy bands take up the extra energy and jump to a higher energy level of the same band. As all the energy levels are delocalised, the electrical energy is easily conducted throughout the metal.

The conductance of metal decreases with an increase in temperature, this is due to the fact that vibratory motion of the atoms increases with temperature which interferes with the free movement of electrons.

(b) Insulator. In an insulator all low energy bands are completely occupied. When the band is completely filled up, electrons fail to conduct electricity, as sufficient energy is not available to them to cross the forbidden zone and jump into the next empty band of higher energy.

(c) Semi-conductor. A semiconductor is that material whose electrical properties lie between those of the insulator and good conductor.

(*a*) An *intrinsic semiconductor,* like Si, Ge, or Se, is a material which is poor conductor of electricity at ordinary temperatures, but the conductivity increases with the increase in temperature.

This type of property arises when the energy band is completely filled up but is separated from the next higher energy empty band by a narrow forbidden zone. At low temperature, the energy is not sufficient to cross the forbidden zone and the material remains

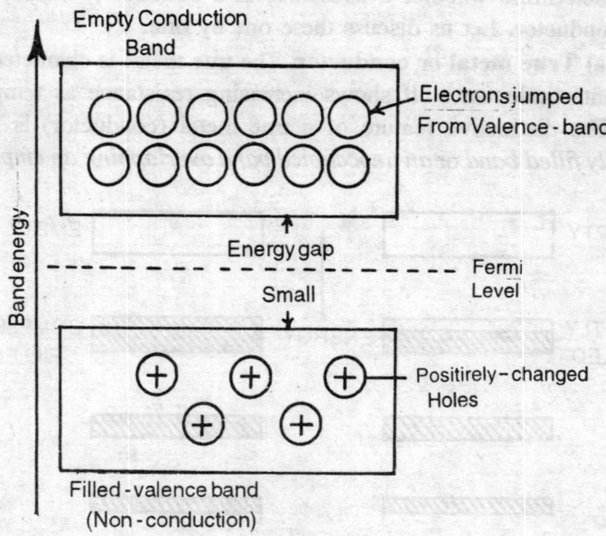

Fig. 4.11. Energy band diagram for an intrinsic semi-conductor at room temperature.

a poor conductor or insulator. However, at increased temperatures the energy of electrons becomes sufficient to cross the forbidden zone with sufficient ease so as to occupy vacant bands, and, thus, making the substance capable of conducting electricity (Fig. 4.11).

(*b*) Substances which are normally insulators can sometimes become semi-conductors when small amounts of impurities are added to it. Such substances are called *extrinsic semi-conductors.* These are of two types:

(*i*) *n-type extrinsic semi-conductor.* This type contains an impurity which has more valency electrons than the parent insulator. This type is obtained by the addition of As or P, pure silicon or germanium (Fig. 4.11).

In this type of semi-conductors, the electron conduction is possible at higher temperatures because the filled impurity levels of As or P are located quite close to the conduction band of Si or Ge.

(*ii*) *p-type extrinsic semi-conductor.* In this type, an impurity is added to an insulator which has less valency electrons than the parent insulator. This type is obtained by the addition of boron or gallium to solid silicon or germanium.

This type of semi-conductor is characterised by the fact that energy levels of B or Ga are quite close to the silicon or germanium valency band. When excited by increasing temperature, electrons from silicon or germanium are excited to the empty impurity level of B or Al and thus leaving positive holes behind. As electrons move into these positive holes, they appear to migrate. Thus, the conduction is due to the movement of positive holes and hence the name *p*-type semi-conductor has been assigned. (Fig. 4.12).

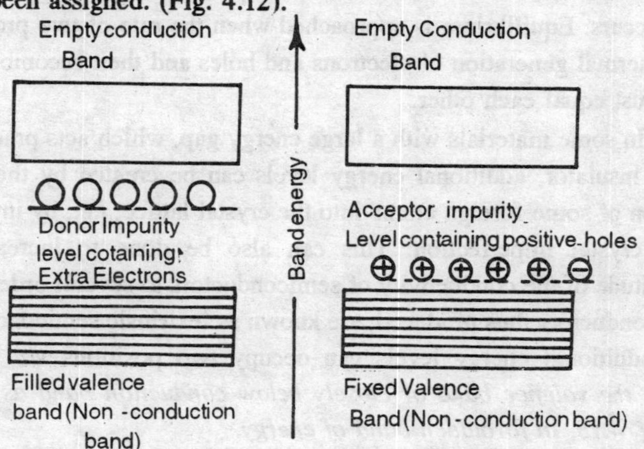

Fig. 4.12. Energy levels for (a) n-type and p-type extrinsic conductor.

4.2. Semiconductors

In semiconducting materials at absolute zero temperature, the so called valence band is the highest fully occupied band. The next band above the conduction band, is totally empty. Between these two energy bands there exists an energy gap known as 'forbidden band'. The forbidden bands in the case of semi-conductors are so narrow that an appreciable number of electrons can easily be raised by thermal excitation, from valence band to conduction band, at the room temperature. The thermal excitation and the transition of electrons into conduction band produces vacancies in the valence band. These vacancies are termed 'holes' and bear unit positive charge. Both, the electrons in conduction band and holes in valence band are mobile and contribute to electrical conductivity. Depending upon the magnitude of electrical conductivity at room temperature the semiconductor is defined as the materials having specific resistance within the range of 10^{-2} 10^{9} ohm-cm.

A semiconductor with a sufficiently high electrical conductivity, produced by the thermal excitation of the electrons is called 'intrinsic semiconductor'. The number of electrons and holes in such a semiconductor equals each other. If an electron jumps from valence band an additional hole is created, thus, positive holes move in opposite direction in valence band to that of movement of electron in conduction band. Some times a free electron meets a positive hole and recombination occurs. Equilibrium is approached when the rate of two processes *viz.*, thermal generation of electrons and holes and their recombination both just equal each other.

In some materials with a large energy gap, which acts practically as an insulator, additional energy levels can be created by the introduction of some foreign atoms into the crystal lattice, *i.e.*, by introducing crystal imperfection. This can also be done to increase the magnitude of the conductivity of semiconductors to several orders. The semiconductors thus produced, are known as 'extrinsic semiconductor'. The additional energy levels can occupy two positions *viz. closely above the valence band or closely below conduction band as shown in Fig. 4.13, in forbidden band of energy.*

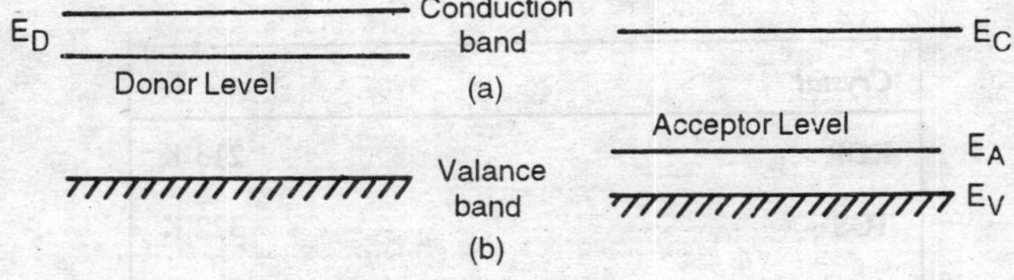

Fig. 4.13.

(a) Structure of energy levels in an *n*-type semiconductor. (b) Structure of the energy levels in *p*-type semi-conductor.

At temperature other than absolute zero, the thermal ionizations of foreign atoms occurs with appreciable probability. When electrons are passed thermally into conduction band, the foreign atoms supplying free electrons are known as donors, and semiconductor described is called extrinsic semiconductor with electron conductivity or *n*-type semi-conductor. Conversely should the foreign atoms have a high affinity for electrons, the electrons from the filled band are transferred to foreign atom, creating holes in valence band. The atoms, which trap electrons from valance band are called 'acceptors' and the semiconductor thus produced is known as extrinsic semiconductor with hole conductivity or *p*-type semiconductor. The most common donors are phosphorous, arsenic, antimony *i.e.* Vth group elements and acceptors are IIIrd group elements namely, boron, aluminium, gallium and indium.

4.3. Ferroelectrics

A ferroelectric crystal shows spontaneous polarisation even in the absence of applied electric field. In the ferroelectric state the centre of the positive and the negative charges of the crystal do not coincide and hence it has a permanent dipole moment. Barium Titanate ($BaTiO_3$,, Tri Glycine sulphate, Rochelle salt ($NaKC_4H_4O_6 . 4H_2O$) Potassium dihydrogen phosphate (KDP) are examples of ferroelectric crystals. The ferroelectric crystals lack a centre of inversion and have very high dielectric constant. On application of cyclic electric fields they exhibit a closed electric hysteresis loop. Above a particular transition temperature, T_C, the ferroelectric crystals loose their ferroelectric properties. Transition temperature for some common ferroelectric crystals is given below:

Crystal	T_c
KDP	213°K
TGS	322°K
$BaTiO_3$	393°K
$LiNbO_3$	1470°K

4.4. Metallic Bonding and Cohesive Energy of the Metals

Metals are characterised by high electrical conductivity and they contain a large number of free electrons. The most weakly bound electrons of the constituent atoms freely move about and are called the conduction electrons. Although there is some contribution of the interaction of the ionic cores with the conduction electrons to the binding energy, the main characteristic of the metallic bonding is a reduction in the kinetic energy of the valence electrons. The wave function for the valence electrons for a metallic crystal as well as a free atom is shown in the Fig. 4.14.

It is seen that the wave function is not changed near the cores but for the outer electrons there is a considerable difference in the metallic crystal. Since $d\psi/dr$ is small for the flat region of the wave functions of the crystal the kinetic energy which is proportional to

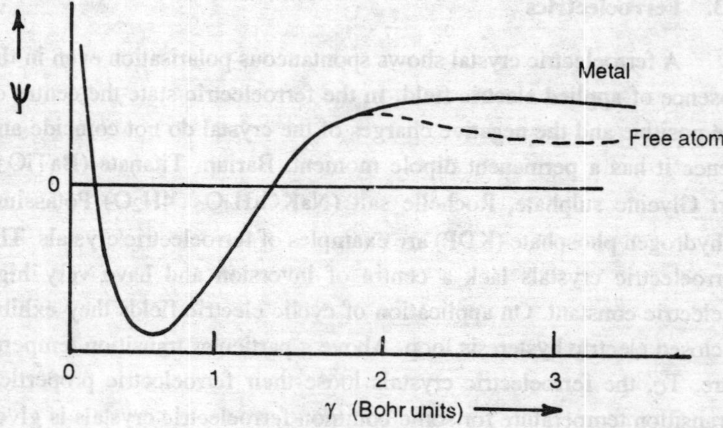

Fig. 4.14. Wave function for valence electrons.

$(d\psi/dr)^2$ is lower for the metallic crystals. In case of sodium it has been estimated that the metal is stable by 1.13 eV with respect to the free atom. This difference between the free atom energy and the crystal energy is called the "cohesive energy".

4.5. Free Electron Theory of Metals

Soon after the discovery of the electrons, Drude and Lorentz tried to explain the high electrical and thermal conductivity of the metals in terms of the free electron theory which was further developed quantum mechanically by Sommerfeld. The free electron theory assumes that the metal contains a large number of free electrons. The free electrons are supposed to be able to move through the lattice in the field of all the nuclei and all other electrons. It is also assumed that the potential is constant everywhere inside the metal. The potential energy of the electrons is lower inside than outside and the change is abrupt at the surface.

First let us consider a particle moving in a one-dimensional potential box. The Schrodinger Equation for the electron is

$$-\frac{h^2}{2m}\cdot\frac{d^2\psi}{dx^2} = E_n\psi_n \quad \text{...(1)}$$

where E_n is the energy of the electron and the wave function has the boundary conditions

$$\psi(x) = 0 \text{ for } x = 0 \text{ and } x = L \quad \text{...(2)}$$

The Schrodinger equation has the solution

$$\psi_n = A \text{ Sin } (n\pi x/L)$$

where A is a constant.

We have

$$d\psi_n/dx = A(n\pi/L) \,.\, Cos\,(n\pi x/L) \quad \text{...(3)}$$

and $$d^2\psi n/dx = -A(n\pi/L)^2 \,.\, \sin(n\pi x/L). \quad \text{...(4)}$$

Hence from the Schrodinger equation we have

$$E_n = (h^2/2m)(n\pi/L)^2 \quad \text{...(5)}$$

Thus for each value of ψ_n there is a wave function with E_n. Evidently the energy spectrum has discrete levels. For an electron in a three dimensional cubic potential box of edge L the energy levels are given by

$$E_{nx}n_yn_z = (h^2p^2/2mL^2)\,(n_x^2 + n_y^2 + n_z^2) \quad \text{...(6)}$$

The parabolic variation of the energy as a function of n is shown in the Fig. 4.15. Though it is shown as a continuous curve, it is actually discreet corresponding to the discrete values of En. Since the energy levels differ by a very small amount of the order of 10^{-18} eV, it can be shown as quasicontinuous.

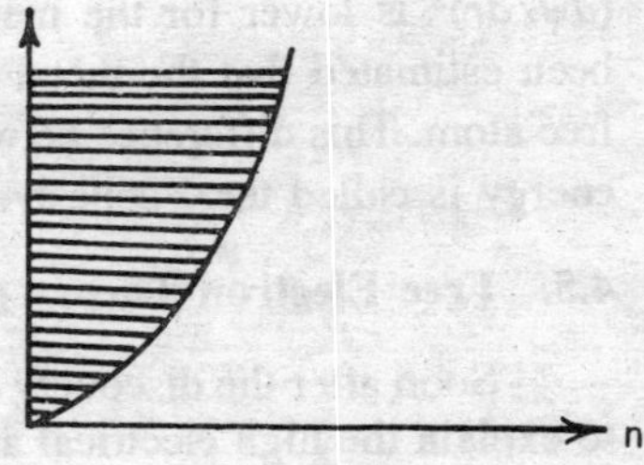

Fig. 4.15. Variation of energy as a function of *n*

4.7. Electron Energies in the Metal

The total energy of the electron as shown in given by

$$E = (h^2\pi^2/2mL^2)(n_x^2 + n_y^2 + n_z^2)$$

Since the energy can be given in terms of the momentum, p by the relation

$$E = p^2/2m \quad ...(7)$$

we have

$$p^2L^2/h^2\pi^2 = n_x^2 + n_y^2 + n_z^2 = R^2 \quad ...(8)$$

One can define a cubic space both as having lattice constant of unity n_x, n_y, n_z being the coordinates.) Thus each cell has unit volume and can have 2 electrons of opposite spin due to Paulis exclusion principle and has energies proportional to the vector R.

Fermi-Dirac distribution function F(E) gives the probability that a particular state with energy E is occupied and is given by

$$\{F(E) = 1/(e^{E-E_F/KT} + 1) \quad ...(9)$$

At T = O, we have,

$$F(E) = 1 \text{ for } E < E_F$$

$$= O \text{ for } E > E_F \quad ...(10)$$

Hence at absolute zero all states below E_F are occupied whereas those above E_F are empty. Hence E_F represents the highest occupied energy level at T = 0 and is called as the Fermi energy. At T > 0, the distribution function is as shown by the dotted line in the Fig. 4.16.

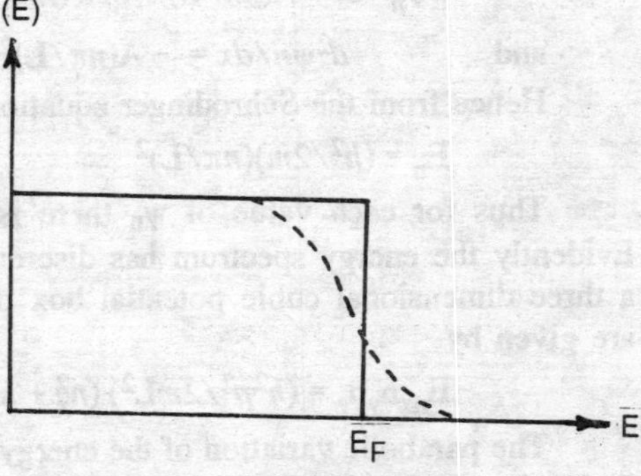

Fig. 4.16. Fermi distribution.

Since each state has the electrons of opposite spin, in a crystal of N electrons there are N/L^3 occupied states. The lattice points lie in the positive octant of the lattice because n_x, n_y and n_z can have positive values only and are enclosed by a sphere of radius

$$\{R_{max} = (2mL^2E_F/\pi^2 h^2)^{1/2} \quad ...(11)$$

Volume of the octant $= 1/8 \times 4/3\pi R^3_{max} = 1/6\pi R^3_{max}$

$$N/2 = 1/6\pi(2mL^2E_F/\pi^2h^2)^{3/2}\} \quad ...(12)$$

or $$E_F = h^2/2m(3\pi^2N/L^3)^{2/3} \quad ...(13)$$

Thus E_F is a function of N/L^3, i.e. the number of electrons per unit volume. The Fermi energies for a number of metals is given below:

Metal	*Li*	*Na*	*K*	*Gu*	*Ag*
E_F	4.72	3.12	2.14	7.04	5.15

(electron volts)

4.7. Zone Theory of Solids

In the free electron theory it was assumed that the free electrons travel in a potential which was assumed constant inside the metal. But actually the potential varies periodically with the periodicity of the lattice. The Schrodinger equation for such a periodic potential can be solved. The interaction of the electron (which can be considered to be a wave of de Broglie wave length λ) with the periodic lattice can also be explained by considering its diffraction by the periodic lattice.

The well-known Bragg diffraction law is given by

$$2\,a \sin = n\lambda \quad ...(14)$$

In terms of the wave vector $K(K = 2\pi/\lambda)$ we have,

$$K = n\pi/a \sin\theta \quad ...(15)$$

or $$K_x = K \sin\theta = n\pi/a \quad ...(16)$$

So when the above condition is satisfied there will be Bragg diffraction. Hence an electron can have different energy states, but the energy state corresponding to $k_x = n\pi/a$ does not exist because the electron suffers diffraction. The portion of the band structure is shown below (Fig. 4.17) for a free electron and an electron with band gap at $k = \pm\pi/a$.

Thus energy discontinuities occur when the wave number k satisfies the critical condition $K = n\pi/a$ with $n = \pm1, \pm2, \pm3....$ Thus the

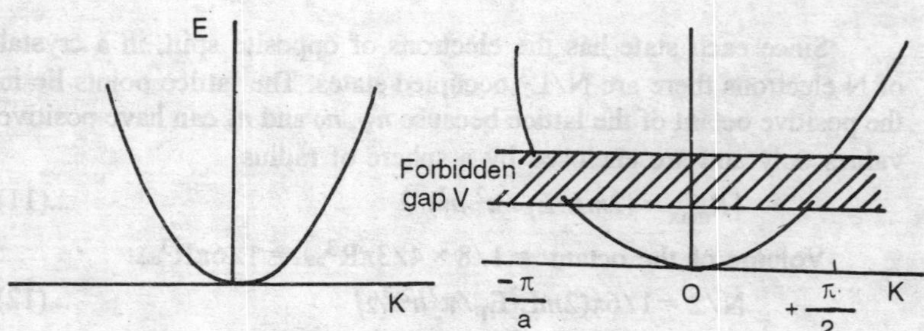

Fig. 4.17. Band structures.

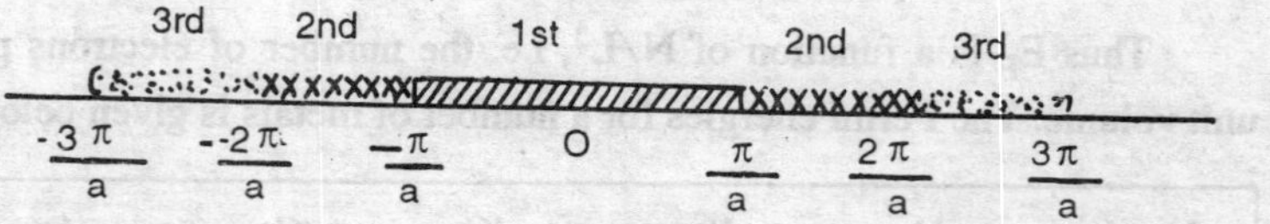

3rd 2nd 1st 2nd 3rd

- 3 π/a - 2 π/a − π/a O π/a 2 π/a 3π/a

Fig. 4.18. Brillouin zones for one dimensional lattice

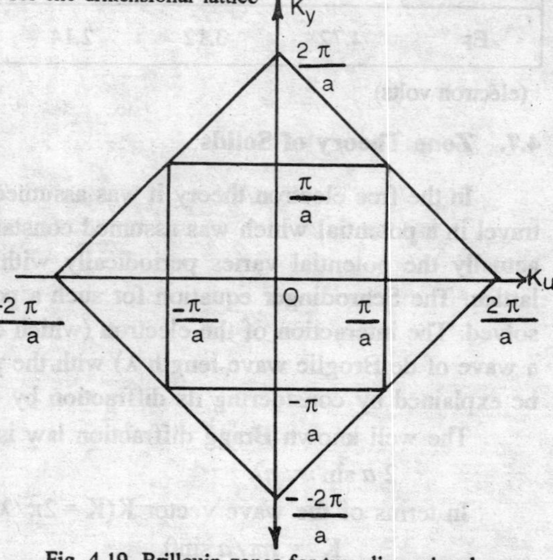

Fig. 4.19. Brillouin zones for two dimensional square lattice.

values of the wave number determines the energy gaps or the forbidden regions. In one dimensional monoatomic lattice the *k*-line is divided into segments each of length π/a. These segments are known as Brillouin zones. The first Brillouin zone is given by $-\pi/a < K < \pi/a$, the second zone is given by the two portions $-2\pi/a < K < -\pi/a$, $\pi/a < K < 2\pi/a$ and so on.

Similarly the Brillouin zones for a two dimensional square lattice is as shown in Fig. 4.19.

The Brillouin zone carries the meaning that E increases continuously with k as long as the value of k is within a particular zone. But when k touches the surface of a particular zone there is a jump in the energy of the electron. The energy gap between two allowed energy zones is known as the forbidden gap.

4.8. Zones in Conductors and Insulators

Let us consider a crystal with N atoms having interatomic separation "a" which is very large. Since the separation is large the interaction can be ignored and they can be treated as free. The electrons in the free atom have their energy levels as per the atomic theory. But when the separation "a" is decreased there is a stronger interaction between them. Since as per the Pauli's exclusion principle no two levels in a crystal can have the same energy, to accommodate the electrons of all the atoms each level splits and forms a band. Thus is level splits to Is band with N levels containing 2N electrons, 2*p* level splits to a band with 3N levels containing 6N electrons etc. The energy separation between the levels in a band is of the order of 10^{-18} eV.

The extent of splitting depends upon the amount of overlap of the orbitals of the neighbouring atoms. When two atoms approach, the outermost levels split first. Thus the extent of split or the amount of smearing of the band is maximum for the outermost band and the smearing is least for the innermost band.

Thus there is an entire band of allowed energy levels corresponding to each allowed energy level in a crystal. Allowed bands alternate with forbidden bands. Electrons cannot have energies in the forbidden band. The higher allowed bands are broader whereas the higher forbidden bands are narrower.

During filling up of the bands the lowest energy states are first

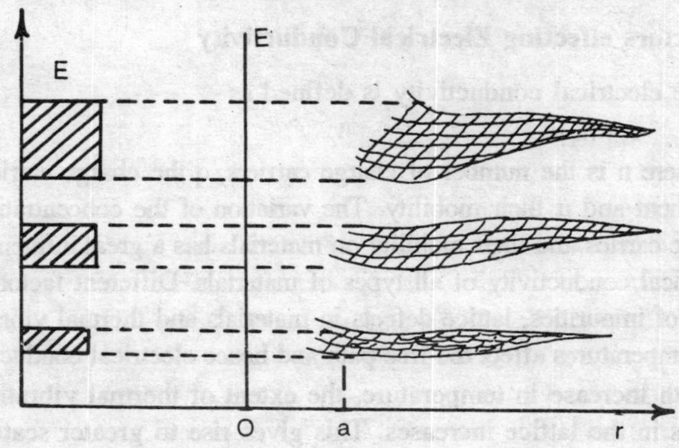

Fig. 4.20 Energy bands

filled up, then the next lowest till all the electrons occupy the energy states. The highest filled state is termed as the Fermi energy. The innermost or the valence band, if it is partially filled, is termed as the conduction band. On application of an electric field the electrons get accelerated and hence attain higher energy. So they have to occupy nearby higher energy levels. If a band is completely filled up, then, since the electrons can not go to the nearby higher levels of the band due to Pauli's exclusion principle, they do not accelerate and participate in the conduction. Hence completely filled or totally empty bands do not participate in the conduction process.

The band structure scheme for metals, insulators and semiconductors is shown in Fig. 4.21.

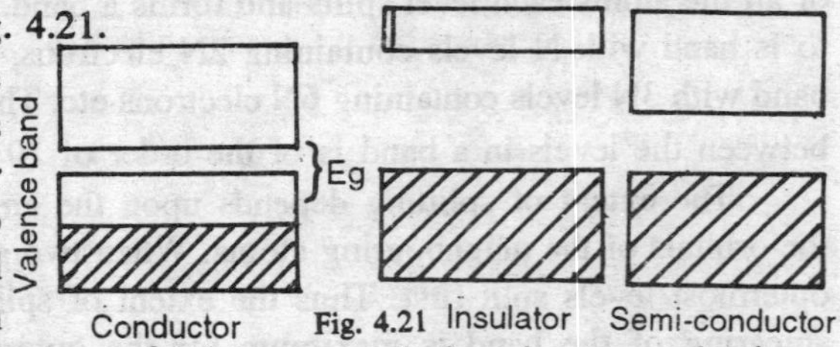

Valence band
Eg
Conductor
Insulator
Semi-conductor

Fig. 4.21

In metals, the valence band is either empty or there is an overlap of the valence and the conduction bands. So the electrons can freely move and give rise to conduction. In case of the insulators the forbidden gap, E_g, is very high and the valence band is totally filled. Hence there is no conduction. In case of the semiconductors the gap is small (2 or 3 eV) and at ordinary temperatures some of the electrons cross over to the higher empty band because of the thermal energy. This creates a few vacant sites on the valence bands known as "holes". On application of electric field, both the electrons and holes give rise to conduction.

4.9. Factors effecting Electrical Conductivity

The electrical conductivity is defined as

$$e = nq\mu \qquad ...(1)$$

Where n is the number of charge carriers, q the charge carried by each of them and μ their mobility. The variation of the concentration of the charge carries and their mobility on materials has a great influence on the electrical conductivity of all types of materials. Different factors like presence of impurities, lattice defects in materials and thermal vibrations at high temperatures affect the free path and hence electrical conductivity.

With increase in temperature, the extent of thermal vibration of the atoms in the lattice increases. This gives rise to greater scattering of the electrons and a decrease in the conductivity. The variation of resistance of the conductors with temperature is linear and is given by

$$p_t = p_o(1 + \alpha\Delta t) \qquad ...(2)$$

Where ρ_t and ρ_0 are the resistivities at temperatures t °C and 0°C, respectively. Δt is the increase in temperature and α is the temperature coefficient of resistivity per degree centrigrade. However, in case of semiconductors, with increase in temperature, more electrons pass on to the conduction band, hence there is an increase in conductivity and a decrease in resistivity.

Presence of impurities in a matrix causes drastic changes in the resistivity due to the scattering by the impurities. There is a reduction in the conductivity and, for dilute concentrations, the resistivity is proportional to the concentration of the impurities.

Due to cold working, filing etc. a large concentration of lattice defects like vacancies, interstitials, dislocations, grain boundaries, stacking faults etc. are produced. These scatter the electrons and thus give rise to changes in the conductivities. Sometimes the faults act as potential barriers to the conduction electrons.

4.10. Conductivity of Semiconductors

As mentioned earlier the conductivity of metals depends upon the presence of free electrons in the crystal due to the metallic bond. In case of semiconductors, the conductivity is due to charge carriers, which depend on many factors like the purity of the semiconductor, temperature etc. The semiconductors can be classified as intrinsic and extrinsic (impurity) or doped. The doped semiconductors can be further sub-divided to two categories as electron or n-type and hole or p-type semiconductors.

4.11. Intrinsic Semiconductors

Intrinsic semiconductors are those that are extremely pure. At temperatures close to absolute zero the atoms of the crystal are covalent bonded and application of field does not cause any directional motion of the electrons. As the temperature increases, some of the electrons attain kinetic energies greater than the binding energy for the covalent bond. These electrons rupture and escape into the interstices of the lattice and become free. These electrons move freely giving rise to conduction. Each electron which moves into the interstitial space becomes a conduction electron. Each electron which moves out of the bond leaves a vacancy or hole. This vacancy gets occupied by an electron from the neighbouring bond, thus creating a hole there. In this way a hole moves in a direction opposite to that of the electron. Since each hole is an electron deficiency it is ascribed a positive charge +e. Thus, in an intrinsic semiconductor the number of holes is equal to the number of electrons and the conduction is due to the movement of both.

The electrical conductivity has two terms and is given by

$$\sigma = en_i\mu_n + ep_i\mu_p$$

where n_i and p_i are concentration of electrons and holes, μ_n and μ_p are the mobilities for the electrons and holes respectively. Since the mobility of the electrons is larger than the mobility of holes the contribution of electrons to the current is much larger.

4.12. Doped (Impurity) Semiconductors

Impurity atoms have great influence on the electric conductivity of semiconductor materials. Conduction due to the electrons or holes of impurity atoms added to the semiconductor is known as the extrinsic semiconduction. There are two different types of impurities which when added to the Ge or Si semiconductor matrix produce excess of electrons or holes. These are called n-type or p-type semiconductors.

4.13. n-type Crystals

Germanium and silicon belong to Group IV in the periodic table and have four valence electrons. The other electrons are lightly bound to the nucleus. These valence electrons form covalent bonds with the four nearest neighbours. When a pentavalent impurity (say As) atom is added to the tetravalent (Ge) matrix, only four of the five electrons of arsenic participate in the covalent bond formation with the four neighbours. The fifth electron does not participate in the covalent bond formation. Its bondage to the nucleus is also weakened because of high dielectric constant of Ge ($\varepsilon = 16$) and the energy of its bonding is some 250 times less. At slightly elevated temperatures KT (KT ~ 0.01eV), the fifth electron gets detached. The semiconductor acquires the conductivity due to free electrons. Since these electrons were not participating in the bond formation, the positive As^+ ions does not behave as holes. They behave simply as positive ions fixed in the lattice and make no contribution to conduction. Since electron conductivity is the dominant factor for conduction in a crystal with the pentavalent impurity (donor impurity), such crystal is called n-type semiconductor. At elevated temperatures the intrinsic electrons and holes of the matrix also contribute to the conduction. The charge carriers whose concentration dominates the conduction process are called majority carriers and the charge carriers of the opposite sign are called the minority carriers.

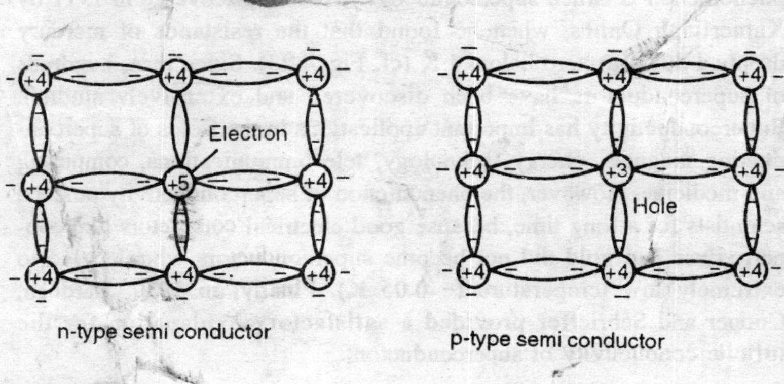

Fig. 4.22 (*a*) & (*b*)

4.14. p-type Crystals

Let us consider the case of a tetravalent germanium matrix containing a trivalent impurity like indium. The indium atom lacks one electron to form four covalent bonds with the four germanium neighbours. The resultant vacancy in the fourth bond represents a hole. The trivalent impurities make available positive carrier or holes that can accept electrons. These impurities are called acceptors. Since the predominant conduction is due to holes, these crystals are called p-type semiconductors. As temperature rises an electron from the neighbouring Ge-Ge covalent bond goes and fills this vacancy, thus making a negative indium ion bonded to the system with a hole in the germanium matrix.

4.15. Insulator

As insulator is a material almost same type as an intrinsic semiconductors. But the bonds are more tightly held by ionic and covalent bonds and much more energy is required to break the electrons from the bonds. The probability of removal of an electron is very small. At extremely high temperatures a few electrons may come out and give rise to conduction. Roughly the conduction is about 10^{12} times less than in case of a good conductor.

4.16. Superconductivity

Certain metals and alloys exhibit almost zero resistivity, i.e., infinite conductivity when cooled to sufficiently low temperatures. This phenomenon is called superconductivity. It was discovered in 1911 by Kamerlingh Onnes, when he found that the resistance of mercury dropped down to zero below 4 K (cf. Fig. 4.23). Since then, hundreds of superconductors have been discovered and extensively studied. Superconductivity has important applications in the fields of superconducting magnets, energy technology, telecommunications, computing and medicine. However, the phenomenon of superconductivity puzzled scientists for a long time, because good electrical conductors like copper, silver and gold did not become superconductors when cooled to extremely low temperature (~ 0.05 K). Finally, in 1957, Bardeen, Cooper and Schrieffer provided a satisfactory explanation for the infinite conductivity of superconductors.

Bardeen-Cooper-Schrieffer (BCS) Theory

In normal metals, plane electron waves, also called de Broglie waves, are assumed to propagate in the direction of motion of the electrons. The free electron gas in a metal can thus be considered as a superposition of several plane, de Broglie waves. In the absence of

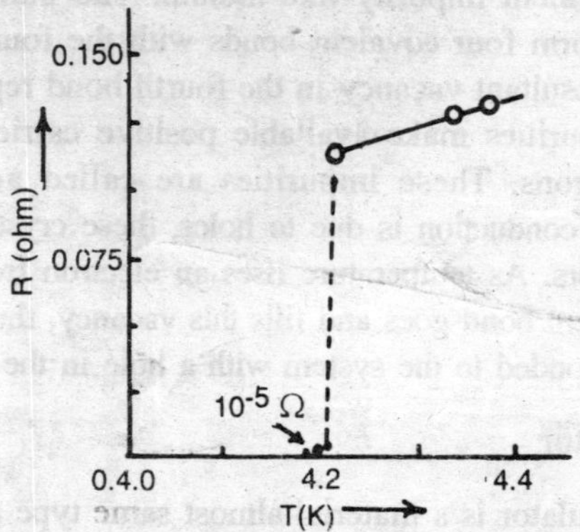

Fig. 4.23 Temperature (*T*) variation of electrical resistance (*R*) of mercury (Hg) indicating transition to the superconducting state

an external electric field, these waves are reflected randomly by the surface of the crystal and there is no net electric current. In the presence of an external electric field, an additional wave component in the direction of the field arises and a finite current starts flowing. For an ideal crystal devoid of any imperfections, there will be zero electrical resistivity (or infinite conductivity). However, in real metals, phonons (lattice waves/vibrations) and defects such as impurities and vacancies act as scattering centres for the de Broglie waves, giving rise to a finite resistance. At high temperatures, phonon scattering plays a major role, while at low temperatures, scattering from static detects predominates.

Bardeen, Cooper and Schrieffer showed that the mechanism of superconductivity is entirely different from that of conductivity in normal, pure metals and is associated with the pairing of conduction electrons into what are known as Cooper pairs. An electron moving through a crystalline lattice draws ions towards it, a situation which attracts another electron. As long as the temperature is low enough, so that the energetically favourable electron pairing is not broken up, the correlation of the conduction electrons gives rise to an almost infinite conductivity. The only requirement is that presence of strong electron-lattice (*i.e.*, electron-phonon) interactions.

A quantum mechanical calculation shows that the free energy is decreased, provided a Copper pair involves electrons with opposite velocities (or momentum) and the lattice serves as the medium for exchange of momentum. In reality, electrons are being continuously exchanged between the Cooper pairs. However, the transition to the superconducting state requires only that a sufficient number of electrons should condense into the paired state. At 0 K, all the conduction electrons are paired, but with increasing temperature, the number of pairs decreases. Anderson showed that non-magnetic impurities do not usually destroy superconductivity. Therefore, this phenomenon could be observed, even for not very pure samples. However, the presence of magnetic impurities lowers the superconducting transition temperature or even suppress superconductivity completely.

A Cooper pair may be considered as a new particle having twice the mass and charge of an electron. In the superconducting state, the electrical current is due to the motion of several Cooper pairs in the same direction with the same velocity; only a single de Broglie wave can then be regarded as collectively representing these Cooper pairs. Since the momentum of the Cooper pairs is small, a given current corresponds to only a small velocity of the wave. Consequently, the wavelength of the de Broglie wave becomes very large, scattering from imperfections rarely occurs, and the conductivity tends to infinity.

4.17. Kirkendall Effect

According to the Graham's law of diffusion and, more rigorously, from the kinetic theory of gases, the rate of diffusion of a gas is inversely proportional to the square root of the molecular mass. As a consequence, when two different gases diffuse into one another, the rate of diffusion is found to be greater for the lighter gas. A similar situation arises when one considers interdiffusion of constituents A and B in a binary alloy AB. The rates of diffusion of A and B are different; usually the lower-melting constituent diffuses faster. The classical experiment by Smigelskas and Kirkendall provided a proof for the existence of vacancy mechanisms of diffusion in pure metals and substitutional alloy, in contrast to rotational or interchange mechanisms. This is now known as the Kirkendall effect. The effect may be understood with reference to Fig. 4.24.

Inert (molybdenum) markers were placed at the welded interface of pure copper and brass (CuZn) pieces as shown in Fig. 4.24. At a high temperature the markers were found to have moved towards one another after a certain interval of time. The distance travelled by the markers was found to be proportional to the square root of the diffusion time. What does this indicate ? If the rates of diffusion of copper and zinc had been equal, the markers would have been stationary. The movement of the markers indicates that zinc atoms diffused faster into the copper side than did copper atoms into the brass. The zinc atoms moved into the vacancies so that the net result was the movement of vacancies, i.e., voids, into the brass side. In other words, extra lattice

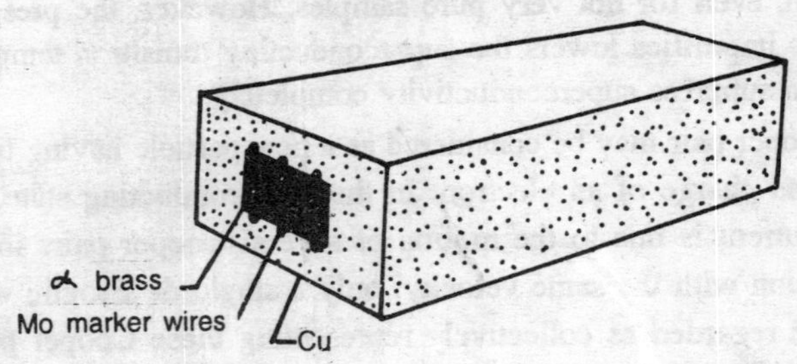

Fig. 4.24. The Kirkendall effect.

planes must have been created on the copper side. Thus, the Kirkendall effect experimentally proves that the vacancy mechanism is operative in metals and substitutional alloys.

5

Mechanical, Thermal, Electrical and Magnetic Properties of Crystals

5.1. Introduction

The physical properties of crystals may be divided into two categories: (1) structurally insensitive, and (2) structurally sensitive. Structurally insensitive properties are those which depend but little, if at all, on a small number of imperfect areas in a crystal. These properties included elasticity, electric conductivity, density, heat conductivity.

The structurally sensitive properties are mechanical strength, internal friction and creep.

Structural imperfections are vacancies in the crystal lattice, foreign atoms, disturbances in the stacking of the layers of material particles, twinning boundaries, and dislocations. A dislocation is understood to be a region in which material particles of a crystal lattice have been moved to different distances from their normal position.

5.2. Mechanical Properties of Crystals

Here we shall discuss those physical properties of crystals which are, along with the optical properties, of great practical importance or can serve as essential diagnostic features. These include primarily mechanical properties: cleavage and hardness.

Cleavage is the ability of crystals to split along definite planes. These planes are always parallel to the actual or possible crystal faces, and their position in space can be designated by the same symbols which are used for designating the positions of faces.

What is cleavage due to ? Material particles making up a crystal are arranged in space at the nodes of a space lattice or several lattices slile one into another. The forces of cohesion between individual material particles depend on the spacings between them. The closer the particles are to each other, the greater are the forces of cohesion, and vice versa, the greater the distance between the particles, the weaker the forces of cohesion.

Thus, we can formulate the following rule : *the direction of cleavage is parallel to the planes with the greatest reticular density.*

For the sake of simplicity, let us first look at the case of plane nets. Fig. 5.1 shows that the denser the rows, the greater the distance between the parallel adjacent rows. For instance, $a_1a_2a_3a_4$ make up a dense row, since the distances $a_1 - a_2$, $a_2 - a_3$, etc., are small. But the rows $a'_1a'_2a'_3$ .... $a''_1a''_2a''_3$. . . parallel to this row are at a rather large distance from it. The row $a''_1a'_4a_7$ is far less dense, but the parallel rows are much closer to it.

In space lattices plane nets function as rows. Where can we expect splitting to be easiest ? Evidently parallel to the faces with the greatest interplanar spacings. In our particular case, if construction is

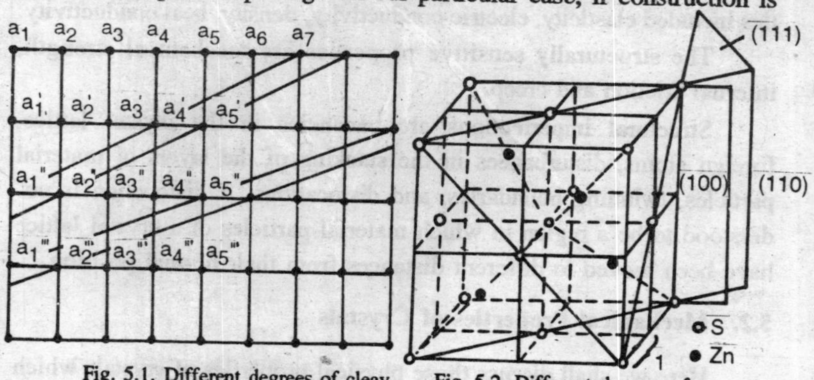

Fig. 5.1. Different degrees of cleavage on a plane net.

Fig. 5.2. Different degrees of cleavage in sphalerite.

carried out on a plane (see Fig. 5.1) splitting will be the easiest parallel to the rows $a_1a_2a_3$ .... $a'_1a'_2a'_3$, etc.

Depending on the structure of the crystal there can be several such directions. Their position and number are directly associated with symmetry.

G.V. Wulff showed that in actual crystals cleavage is more complicated, since chemical bonds are added to the forces of cohesion. For instance, in sphalerite, ZnS, (Fig. 5.2), which belongs to the cubic system, the interplanar spacings in the ionic lattice are: 0.500 for the cube (100), if the length of an edge of small cube (one of the eight cubes into which the unit cell can be broken) is taken to be unity, 0.707 for the rhombic dodecahedron (110), and alternately 0.288 and 0.866 for the octahedron (111).

It would seem here that cleavage would be parallel to the octahedron which has the largest interplanar spacing (0.866). But actually it is parallel to the rhombic dodecahedron where inter-planar spacing is 0.707. This is because the layers perpendicular to the octahedron faces consist alternately of ions of zinc and sulphur, while the layers along the rhombic dodecahedron contain both kinds of ions. In the latter case chemical bonds between adjacent layers are evidently weaker than between layers of the octahedron which consist of ions with different charges (see Fig. 5.2).

The quality of cleavage is evaluated in accordance with the surface of a split piece, and usually five degrees of cleavage are distinguished: (1) eminently perfect, excellent — the crystal can be split into finest laminae with mirror smooth surfaces (mica); (2) perfect—smooth shiny surfaces (calcite); (3) distinct—smooth surfaces and fractures (feldspars); (4) imperfect—almost no smooth planes, uneven fracture (beryl and apatite) in the direction [0001]; (5) poor—cleavage direction is found with great difficulty (quartz, corundum). Crystals of the same substance have several directions of cleavage of similar or different degree of perfection.

Hardness of crystals is of great practical importance, especially for the abrasives industry; it is also an important diagnostic feature helpful in the recognition of minerals. It is one of the properties with respect to which crystals are strongly anisotropic. The evaluation of hardness greatly depends on the method by which it is determined. Below we discuss the principal methods of determination of hardness.

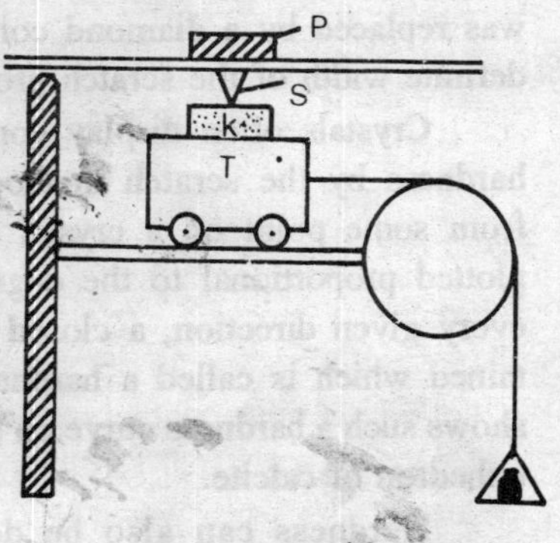

Fig. 5.3. Zeebeck sclerometer.

Scratch method. If one crystal scratches another, *i.e.*, leaves a trace on it, then obviously one is harder than the other. If both crystals scratch one another, they are considered to have equal hardness. The mineralogist Mohs evalued a special scale of ten standard minerals with hardness increasing from one (the softest mineral) to ten (the hardest mineral). Hardness is determined by the number of the mineral in the scale. If a crystal is harder than one of the standards (for instance, harder than 3) and softer than the next number (4), its hardness is expressed by the intermediate value, *i.e.*, $3\frac{1}{2}$. The Mohs scale consists of very widespread minerals (except No. 10):

1. Talc	5. Apatite.	8. Topaz
2. Gypsum	6. Orthoclase	9. Corundum
3. Calcite	7. Quartz	10. Diamond
4. Fluorite		

The quantitative evaluation of hardness by the scratch method is carried out with the aid of special device called a *sclerometer*. The Zeebeck sclerometer is shown diagrammaticaly in Fig. 5.3.

A crystal *k* is placed on a movable block *T*. A steel point *S* is pressed against the crystal by a weight *P*. The block with the crystal is moved under the steel point, and the weight *P* is varied until the point scratches the crystal.

A sclerometer of this type has two important drawbacks: (1) it cannot be used for substances which are harder than the steel of the point; (2) it is very difficult to determine the exact weight at which a scratch begins to appear.

The instrument was later improved by Martens: the steel point was replaced by a diamond cone of a definite angle (90°), so that a definite width of the scratch groove was established (0.01 mm).

Crystals often display considerable anisotropy when tested for hardness by the scratch method. For instance, if from some point on a crystal face segments are plotted proportional to the degree of hardness in every given direction, a closed curve will be obtained which is called a hardness curve. Fig. 5.4 shows such a hardness curve on the cleavage rhombohedron of calcite.

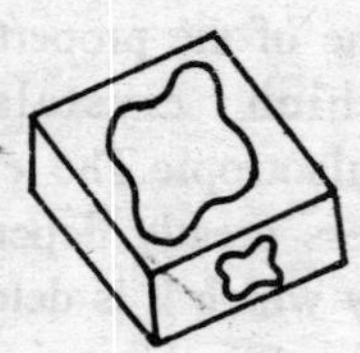

Fig. 5.4. Hardness indentation figure in calcite

Hardness can also be determined by the *grinding method*. A crystal is placed on a cast iron or glass disc and is ground with powedered carborundum, CSi — a substance of great hardness; both the crystal and the powder are previously weighed. Grinding is carried to destruction, which is a rather inaccurate criterion. The hardness of the crystal is determined by the loss of weight.

Minerals are also tested for hardness by an indenting method with the aid of a *microhardness tester,* an instrument designed in 1943 by M.M. Khrushchov and Y.S. Berkovich.

The indenting tool is a diamond pyramid square in section with an angle of 136° between opposite faces. The weight applied varies from 2 to 200 kg. The hardness number is expressed in kilograms per square millimetre — kg/mm^2. Measurement is taken along the indentation diagonal. The instrument is connected to a conventional biological microscope which gives a 400-fold magnification and has a lamp for work in reflected light. The microscope stage is so arranged that after rotation through 180° the object under the objective is placed exactly under the weight terminating in the diamond pyramid.

Hardness tests can be conducted on minerals of any Mohs number.

In Table 5.1 comparative values are given which were obtained by the sclerometer and grinding methods for the ten standard minerals of the Mohs scale.

We can see from Table 5.1 that the increase in the hardness of standard minerals from one number to another is far from uniform.

Table. 5.1

Mineral	*Relative hardness measured with*	
	Sclerometer	*grinding method*
Talc	2.3	0.03
Gypsum	9.5	1.25
Calcite	22.5	4.5
Fluorite	25.5	5
Apatite	35.5	6.5
Orthoclase	108	37
Quartz	300	120
Topaz	450	175
Corundum	1,000	1,000
Diamond	—	140,000

The most insuccessful standards are calcite, fluorite, and apatite which differ but little from each other in hardness. Notable is the hardness of diamond which is 140 times greater than that of its neighbour, corundum.

The surface graph for hardness cannot be drawn in the form of an ellipsoid.

Only two general postulates can be given for the causes of different hardness of crystall bodies.

1. All other conditions being equal, hardness depends on the interplanar spacings — the smaller the spacings, the greater the hardness.

2. Hardness increases with valency. For instance, NaF, MgO, ScN, and TiC have the same structure (of sodium chloride type) and very close interplanar spacings, but their ions are of different valencies. Their hardness is given in the table below.

Chemical composition	NaF	MgO	ScN	TiC
Valency of ions	1	2	3	4
Interplanar spacing in the direction (100), in Å	2.31	2.10	2.23	2.23
Hardness on Mohs scale	3.2	6.5	7.8	8.9

Elasticity. Alteration in the shape of body, caused by some reason or another, is called *deformation.*

External mechanical forces applied to a crystal cause deformations which can be plastic or elastic in nature.

If a force *P* does not exceed a certain limit, the cessation of the action of this force is followed by the disappearance of the deformation. Such a deformation is *elastic.*

Elasticity is that property of solids by virtue of which they recover their former shape and dimensions when external forces cease to act.

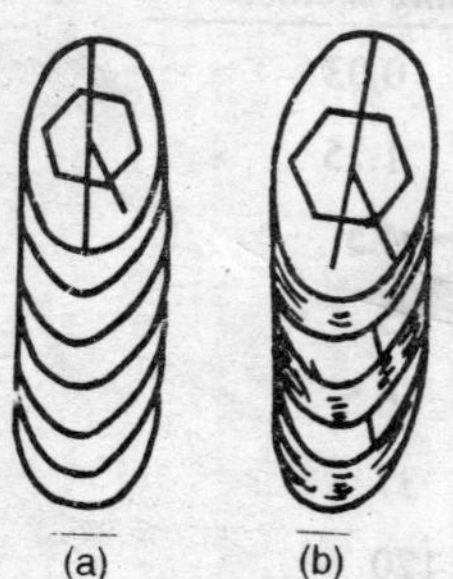

Fig. 5.5. Gliding in a zinc crystal:
a-crystal before deformation;
b-crystal after deformation.

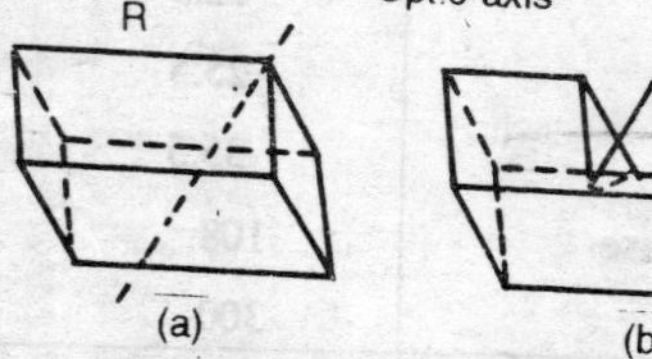

Fig. 5.6. Gliding in a calcite crystal:
a-crystal before deformation;
b-crystal after deformation.

The force P can be increased to a value after which a body will not recover its former dimensions completely, and some deformation will remain. Such a deformation is called *permanent or plastic.*

The *elastic limit* is the maximum stress intensity $\left\{\frac{\sigma}{S}\right\}$ which leaves no permanent deformation.

Of the types of plastic deformation we shall briefly discuss gliding, and mechanical twinning.

Gliding consists of parallel translation of crystal layers along one constant (stationary) plane called the *gliding plane*. The translation of individual layers is directly proportional to their distance from the gliding plane. In gliding, the particles of a crystal retain their mutually parallel orientation.

A crystal of zinc may be used to illustrate the point. A cylindrical crystal (Fig. 5.5) elongated parallel to the axis loses its original shape and becomes somewhat flattened in the direction of gliding (small arrows in Fig. 5.5).

In *mechanical twinning* there is also some translation of layers along the twinning plane, but in this case it obeys some twinning law, and particles change their orientation.

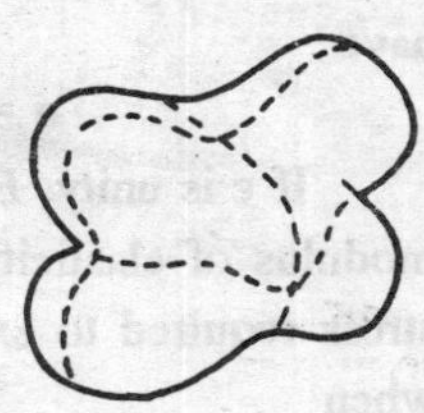

Fig. 5.7. Elasticity surface of barite.

Mechanical twinning is characteristic of many crystals, for example, of calcite. Pressing a knife blade to one of the edges of a cleavage rhombohedron R (Fig. 5.6) will produce twinning of the crystal parallel to the plane of the form (0-12).

When graphically represented, the elastic properties of crystals produce intricate surfaces, as can be seen from Fig. 5.7 showing the surface of elasticity of barite (orthorhombic system).

In 1675 Robert Hooke formulated the following law (which was named after him) for isotropic bodies: *deformation of a solid is proportional to the stress producing it.*

In the case of extension Hooke's law can be expressed mathematically as follows. If the original length of body is called L, its cross-section S, and the elongation produced ΔL, then the elongation per a unit length (relative elongation) will be

$$\frac{\Delta L}{L} = \varepsilon.$$

The tensile stress (*i.e.*, the load per unit cross-sectional area of the rod) is

$$\frac{P}{S} = \sigma.$$

Hooke's law, in the case of extension, will appear as

$$\varepsilon = \sigma e \qquad ...(1)$$

i.e., relative elongation is directly proportional to the tensile stress. The number e is called the coefficient of linear extension (or compression). Its reciprocal $E = \frac{1}{\varepsilon}$ is called the modulus of elasticity or the Young modulus. Rewriting the equation (1) with E we shall have

$$\sigma = Es. \qquad ...(2)$$

If ε is unity, E in this case will be equal to σ_1. It means that the modulus of elasticity of a substance is numerically equivalent to the stress required to extend a rod to double its original length, because when

$$\varepsilon = 1, \quad \text{then} \quad \frac{\Delta L}{L} = 1, \quad \text{and} \quad \Delta L = L.$$

A few words should be said about the so-called *percussion figures* which are sometimes used as diagnostic features in a rough recognition of a mineral.

If the point of a metal rod is placed on a crystal plate and the other end of the rod is struck with a hammer, fractures will appear around the point running always in the same directions depending on the symmetry of the crystal. These fractures are called percussion figures and serve as diagnostic features.

Fig. 5.8. Percussion figure on mica

Fig. 5.8 shows a percussion figure on the face (001) of mica — a monoclinic pseudohexagonal mineral.

Heat conductivity of crystals was first studied by Senarmont in 1847. Different faces of a crystal or plates cut from it were covered with a thin layer of wax. Heat was conducted through a wire which was insulated with asbestops to prevent radiant heat from melting the wax. The wire was wrapped in asbestos, only its free end contacting the crystal through a hole in an asbestos sheet.

The wax melted at the point of contact, and in most cases the melted area was in the shape of an ellipse. If the test was conducted on a pinacoid (001) or (0001) of a crystal belonging to an intermediate system, the melted area was in the shape of a circle.

The heat conductivity properties of different crystals are clearly similar to their optical properties. The heat conductivity surface of optically uniaxial crystals has the form of an ellipsoid of rotation. For optically biaxial crystals, *i.e.*, crystals of the orthorhombic, monoclinic, and triclinic systems, the surface of heat conductivity has the form of an ellipsoid with three unequal axes. It has two cross-sections in which heat conductivity is propagated in a circle, similarly to two circular sections of optically biaxial crystals.

As regards their thermal properties crystals are rather strongly anisotropic. The heat conductivity of some crystals may be 3 to 4 times higher in some directions than in others. It is usually higher than that of amorphous substances of the same chemical composition. This difference decreases with an increase in temperature.

Eucken has established the following laws governing heat conductivity.

1. Crystals usually have a higher heat conductivity than amorphous substances of the same composition.

2. The heat conductivity of crystals is approximately inversely proportional to absolute temperature (*T*). The heat conductivity of amorphous bodies decreases with decreasing absolute temperature.

3. The more complex the chemical composition of a crystal, the lower its heat conductivity.

Apart from this, Jeannetta arrived at the following empirical conclusion: the heat conductivity of crystals is much higher in the directions parallel to cleavage, than in those perpendicular to it. The data of table 5.2 illustrate the above statements.

The coefficient of thermal expansion of crystals also varies with direction, sometimes rather abruptly. In some very rare cases even compression is observed during heating, *e.g.*, in calcite, in the directions perpendicular to the principal axis.

Table 5.2
Coefficient of Heat Conductivity, cal/cm sec deg.

Temperature absolute / Celslus	Substance: quartz, parallel to cleavage	quartz, perpendicular to cleavage	Quartz glass
373/100	0.021	0.013	0.0046
273/0	0.032	0.017	0.0033
195/-78	0.047	0.024	0.0028
83/-190	0.117	0.059	0.0016

5.3. Electrical Properties of Crystals

The *electrical conductivity* of crystals, in contrast to that of amorphous bodies, varies with direction. There are two kinds of electrical conductivity — *ionic conductivity,* which is inherent in solutions, and *electronic conductivity* (in metals).

Specific resistance of quartz (a dielectric) in the direction of the principal axis is hundreds of times lower than in the directions perpendicular to it. Resistance increases considerably with an increase in temperature. The specific resistance of metals also varies with direction, but the variations are negligible. The increase of temperature has less effect on metals than on non-conducting crystals.

Of great interest are the *pyroelectric* and *piezoelectric* properties of crystals. Recently these properties have acquired a considerable industrial importance.

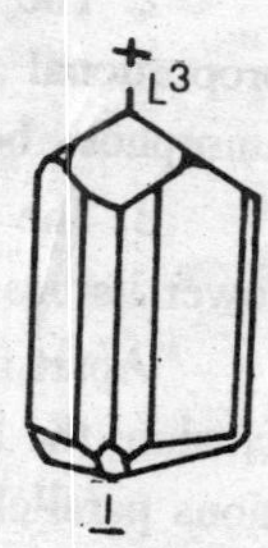

Fig. 5.9. Positive and negative charges in a tourmaline crystal.

Pyroelectricity consists in the following. If a crystal is transferred into an environment with a temperature differing from that in which it was kept prior to the test, some of its faces become electrically charged. This phenomenon is observed only in dielectrics, of course. It was first discovered in tourmaline which has the

$L^3 3P$ symmetry (Fig. 5.9). The pole which develops a positive charge with an increase in temperature is called the *analogous pole (+)*, the opposite one being the *antilogous pole (–)*.

Pyroelectrical phenomena can be investigated by several methods. The simplest is the dusting method. A crystal is dusted with a finely powdered mixture of equal volumes of flowers of sulphur and red lead. In their passage through the silk sieve of a pulverizer the particles of powder become charged by friction, the sulphur receiving a negati[illegible] [illegible]harge, and the red lead a positive one. The particles are attracted and retained by those parts of the crystal which have the opposite charge. The distinct difference in the colouring of particles enables us to detect easily the presence of pyroelectrical phenomena on tourmaline crystals.

The quantity of pyroelectricity can be measured with the aid of a very sensitive electrometer.

Pyroelectricity is displayed only by crystals with *special* polar axes, i.e., such axes which have no congruent or reflection equivalents. *Polar* axes are axes whose opposite ends cannot be placed in a congruent position by any element of symmetry of the crystal. Special polar axes, and consequently pyroelectrical phenomena, are possible only in the following ten crystal classes:

$L^1, L^2, L^3, L^4, L^6, P, L^2 2P, L^3 3P, L^4 4P, L^6 6P$.

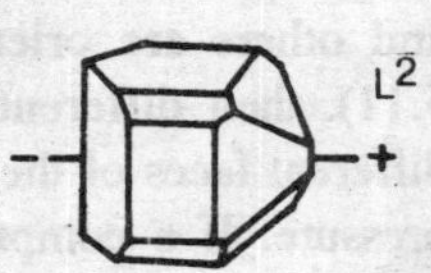

Fig. 5.10. Positive and negative charges in a tartaric acid crystal

Pyroelectricity is very pronounced in crystals of tourmaline belonging to the $L^3 3P$ class (Fig. 5.9) and of tartaric acid belonging to the L^2 class (Fig. 5.10).

Electrical charges may also develop in some parts of crystals of dielectrics subjected to mechanical pressures, which may be either positive (extension) or negative (compression). In both cases opposite charges will develop at the same spot of the crystal. This is called a *direct piezoelectric effect*. It has also been found that in an electric field the crystal will suffer elastic deformation, i.e., will be extended or compressed. This is called a *reverse piezoelectric effect*.

The charges develop at the ends of axes which have a polar development. Therefore, piezoelectric effects, both direct and reverse, can appear only in crystals without a centre of symmetry, i.e., in those which belong to one of the following calsses: $L1, L^2, L^3, L^4, L^6, L^2_4, P, L^22P, L^33P, L^44P, L^66P, L^3P, 3L^2, L^33L^2,$ $L^4 4L^2, L^66L^2, L^2_4 2L^22P, L^33L^24P, 3L^24L^3, 3L^44L^36L^2, 3L^24L^36P.$ The last class but one does not count because all piezoelectric constants corresponding to it are zero.

Both the direct and the reverse piezoelectric effects have been widely applied in science and engineering, principally in the measurement of pressure in a gun barrel during discharge, wavelength stabilization in radio transmitters, sounding of sea depths, transmission of directed radio signals, *i.e.*, signals which can be received only by the particular receiver to which they are sent. Quartz and rochelle salt are the principal piezoelectric crystals used.

The intensity of the piezoelectric effect in a crystal depends on the crystallographic direction. Moreover, crystals of different substances produce effects widely differing in intensity.

Let us discuss the following example. All the three twofold axes of quartz (symmetry L^3L^2) have polar development. If a plate in the form of a parallelopiped is cut from a quartz crystal so that some faces are perpendicular to one of the L^2 axes, which are called electric axes, and others are oriented parallel to L^3 (Fig. 5.11). then different charges will develop at different faces of the plate upon application of pressure. If a compression force f is applied parallel to L^2, then one of the faces normal to it will receive a positive charge, and the other a negative one.

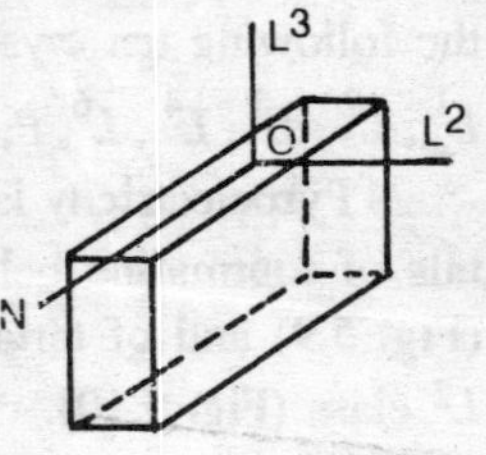

Fig. 5.11. A quartz plae cut perpendicularly to the electric axis

The quantity of electricity, q, does not depend on the size of the plate, *i.e.*, $qL^2 = kf$, where k is the piezoelectric constant of quartz. If the compression force f is applied parallel to L^3 no charges will develop on the faces of the plate: $qL^3 = 0$. If the compression force f is perpendicular to L^2 and L^3 (N in Fig. 5.11), opposite charges will develop on the same faces as in the first case: $q_N = k\frac{n}{b}f$, where n is the size of that edge of the plate along which the force facts, and b is the thickness of the plate in the direction of the electric axis L^2.

5.4. Magnetic Properties of Crystals

In order to determine whether a substance is paramagnetic or diamagnetic it is necessary to disturb the orientation of its particles. The substance is crushed into fine powder which is placed in a glass capillary suspended by a thread in a magnetic field.

Induced para- or diamagnetism is equally strong in all directions only in singly refracting crystals. A ball made from such a crystal will occupy no definite position in a magnetic field.

Induction along the optic axis of optically uniaxial crystals will be either maximum or minimum. A ball made from a paramagnetic crystal should be so suspended that it can revolve about one of the diameters perpendicular to the optic axis. If the latter is the direction of the maximum magnetic induction, the ball will take up a position parallel to the lines of force, and conversely, if the axis coincides with the direction of the minimum induction, the ball will take up a position perpendicular to the lines of force. A ball made from a diamagnetic crystal will behave in the opposite manner.

In optically biaxial crystals the value of magnetic induction corresponds to an ellipsoid with three unequal axes. If a crystal has three elements of symmetry, the directions of magnetic induction will coincide with the principal directions of light vibrations. In crystals with one symmetry element only one direction is oriented parallel to the principal direction of the vibrations of light. In crystals of the triclinic system there is no relationship between the principal directions of magnetic and optical vibrations.

6

Crystallography and its Relationship with other Branches of Science

(A brief outline of the History of Development of Crystallography)

6.1. Introduction

Crystallography is the science of crystals which is devoted to the study of their development and growth, their external, form, internal structure, and physical properties.

The word "crystal" is of Greek origin, it is formed from two words: χρυοζ—cold, and στελλεσθαν—to congeal, and means "congealed by cold". The Greeks, as well as the scholars of later ages (Stensen, 1638-1686), applied the word χρυζταλλον to rock crystal, as they believed it to have been formed from ice crystals. Some Greek philosophers thought that ice would always change to a specific form, stable at normal temperature, if it was kept for some time at extremely low temperature, *i.e.*, if it was "congealed by cold".

For a very long time crystallography was a part of mineralogy, an introduction to it. At the end of the 19th century, however, became an independent discipline for several reasons the most important of which were the following.

1. With the progress of chemistry, especially of organic chemistry, it became evident that tremendous number of substances, having nothing in common with minerals, had a crystalline structure.

2. Crystals, or more precisely, crystalline substances, are very widespread. The overwhelming majority of bodies and objects surrounding us have a crystalline structure. The earth and the stones we

walk on, the walls and roofs of our houses, even wood — all this has a totally or partly crystalline structure, often a very complicated one.

3. Crystals have highly diversified specific properties.

6.2. X-ray Crystal-Structure Analysis

In 1895 Röntgen discovered an unusual type of radiation which he called X-rays, the nature of which remained obscure for a long time.

In 1912 the physicist Laue found that these rays were of an electromagnetic nature. He succeeded in effecting their interference by using a zinc blende crystal (ZnS) as a three-dimensional diffraction grating. Laue's work made it possible to determine the wavelength of X-rays and to measure the spacing between material particles in crystal lattices. These spacings and wavelengths are measured in angstroms (abbreviated to Å), one angstrom being 1×10^{-8} cm, or one ten-millionth part of a millimetre.

X-rays are obtained in tubes from which air is almost completely evacuated. The rays emanate from an anode when it is bombarded by a high velocity stream of electrons. The voltage required runs into thousands and tens of thousands of volts (1,000 volts, or a kilovolt, is designated by the letters kV). Direct current used is of the order of a few milliamperes (mA). X-ray waves, similarly to those of visible light, differ in length — from 0.1 to 10.0 Å.

Short-wave radiation is called *hard* radiation, long-wave radiation is *soft*. The character of radiation, including its intensity and wavelength, depends on the chemical composition of the anode metal and on the voltage of the tube circuit. The rays emanate at a small angle from the surface of the usually flat anode. If the waves are all of one definite length, the radiation is called *monochromatic*. If it contains waves of different length, the radiation is called *white* or *polychromatic*, by analogy with visible light.

To obtain monochromatic radiation, a voltage is selected depending on the metal of the anode, and a filter is placed in the path of the rays to suppress all radiation except that of a definite wavelength. The filters are chosen in accordance with the wavelength of the characteristic radiation. A manganese filter is used with an iron anode, a nickel filter with a copper anode, a zirconium filter with a molybdenum anode. Instead of the pure metals, their oxides or salts can be used for filters. They are introduced into a celluloid film in the amount of 0.0042 g of manganese, 0.0067 g of nickel, or 0.02 g of zirconium per 1 sq centimetre of film. Monochromatic rays used for crystal-structure analysis are usually of considerable wavelength: $\lambda = 1.539Å$ for a copper anode, and $\lambda = 1.934Å$ for an iron anode.

As we see it, the interference of X-rays is caused by the crystal lattice in the following manner. The rays from the anode of a tube, or *primary* rays, impinge on a crystal, whose material particles begin to vibrate and emite rays of the same wavelength but of much lower intensity. These rays are called *secondary*, and as they have a very low intensity their presence can be detected only if they combine by interference and thus in crease their intensity. G.V. Wulff in the U.S.S.R. and W.H. Bragg and W.L. Bragg in Great Britain drew up a diagram and worked out a formula for this phenomenon.

Let us imagine a section through a crystal lattice (Fig. 6.1). Plane nets a_1, a_2, a_3, etc., are normal to the plane of the drawing. Material particles of the crystal are located in the nodes of the nets. A bundle of parallel X-rays (primary rays) strikes this series of planes at an angle θ. The angle is made by the direction of the rays and the plane of incidence (and not the normal to this plane, as in optics). The θ angle is therefore called *glancing angle.*

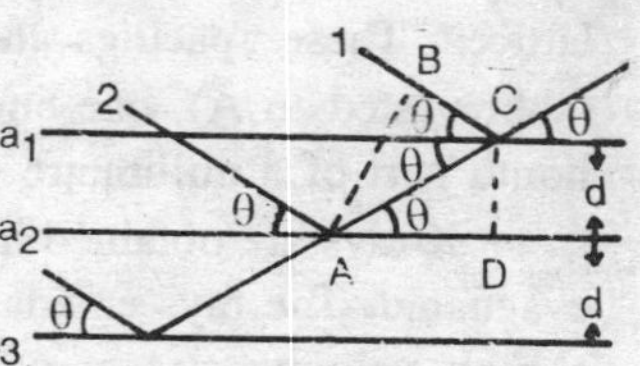

Fig. 6.1. A drawing to the derivation of the Bragg-Wulff equation.

The secondary rays emitted by atoms in the planes a_1, a_2, a_3, etc., interfere, as there is a phase difference, Δ, between them. The interference leads to an increase in intensity, if the phase difference represents a whole number of waves, *i.e.*, if $\Delta = n\lambda$, where n = 1, 2, 3, etc. The phase difference depends on the glancing angle θ and on *d*, the spacing between the plane nets a_1, a_2, a_3, etc. The three values are related to each other in the following way. The phase difference of rays *1* and 2 $\Delta = AC - BC$. From the right triangle *ABC* we have:

$$BC = AC \cos 2\theta,$$

hence,

$$\Delta = AC(1 - \cos 2\theta) = 2AC \sin^2 \theta.$$

From the triangle *ADC* in which $CD = d$, *i.e.*, interplanar spacing, we have :

$$AC = \frac{CD}{\sin\theta} = \frac{d}{\sin\theta},$$

hence,

$$\Delta = n\lambda = \frac{d}{\sin\theta} . \sin^2\theta = 2d\sin\theta$$

This is the Bragg-Wulff formula, the principal formula of X-ray crystal-structure analysis.

The glancing angle θ can be determined experimentally. Knowing the angle and the wavelength λ we can determine the interplanar spacing *d*. If *d* is known, λ can be calculated. In determining θ, λ and *d* monochromatic radiation of constant wavelength is used.

There are four principal methods of crystal-structure analysis.

1. Laue Method. A bundle of X-rays of different wavelengths impinges at 90° on the plane of crystal plate 0.2-1.0 mm thick or on a small crystal. On their way through the crystal the rays encounter plane nets with different spacings *d*. The nets make various angles θ with the direction of the rays. Some combinations of *d*, θ and λ satisfy the conditions of the Wulff-Bragg formula with the resulting increase in radiation intensity.

The Laue method is shown in Fig. 6.2. Primary rays pass through the crystal *k* via the inlet diaphragm *f* and produce a central black spot at the point C on the photographic plate PP. We recall that the bundle consists or rays of different wavelength. The number of plane nets forming various angles θ with the direction of the incident ray is infinitely large. Under these conditions some planes will increase the intensity because of the interference of secondary rays, which show up on the photographic plate as weak spots surrounding the central spot.

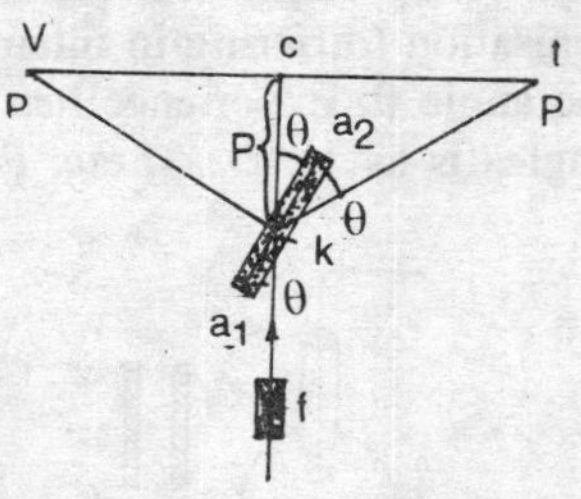

Fig. 6.2. Laue method.

In our example (see Fig. 6.2) a_1a_2 is one of the possible plane nets, therefore, θ is the corresponding glancing angle, *t* is the spot on the photographic plate, *V* is the gnomonic projection of the a_1a_2 plane or of the crystal face corresponding to it.

In a photograph made by this method we can measure the distance from the centre *C* to any spot *t*. The value of *R*, the distance from the crystal to the photographic plate, can also be measured. Then, insofar as $Ct = R \tan 2\theta$ we can determine the θ angle for the corresponding plane, and plot the spot *t* on a gnomonic or stereographic projection.

Thus, the Laue method enables us to investigate the symmetry of the crystal lattice and to determine the orientation of plates which were cut from a crystal at unknown angles to the elements of symmetry of the crystal.

2. Bragg Method. A bundle of monochromatic X-rays impinge on the crystal *k* via the diaphragm *S* (Fig. 6.3). The crystal is mounted on a stage rotating about a vertical axis (i.e., perpendicular to the plane of the drawing). Secondary rays intensified by interference are registered not photographically but with the aid of an ionisation chamber υ. The chamber is a hollow lead cylinder filled with a gas, *e.g.*, SO_2. The gas may be ionized by X-rays. Two electrodes, *l* and *t*, are connected to a storage battery *B*. An electrometer *E* showing the degree of ionisation is connected in the same circuit. Ionisation takes place when the angle θ corresponds to the interplanar spacing *d* of that crystal face on which the primary bundle impinges. To find the value of the angle θ the ionisation chamber is rotated about the axis of rotation of the stage with the crystal. Readings are taken off the dial *L*. Ionisation (differing in intensity) may take place at different values of the angle θ. Experience has shown that the ratio of the sines of these angles is as 1 : 2 : 3, etc. For instance:

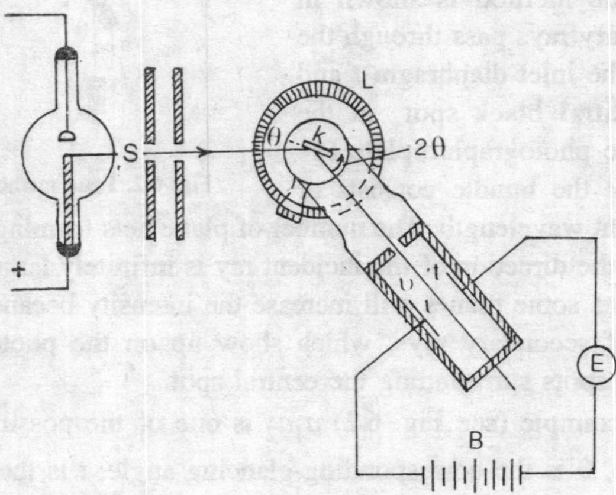

Fig. 6.3. Bragg method.

$$\theta_1 = 10^\circ 40', \quad \sin\theta_1 = 0.185 = 1 \times 0.185 = 1 \sin\theta_1$$

$$\theta_2 = 21^\circ 40', \quad \sin\theta_2 = 0.370 = 2 \times 0.185 = 2 \sin\theta_1$$

$$\theta_3 = 33^\circ 40', \quad \sin\theta_3 = 0.555 = 3 \times 0.185 = 3 \sin\theta_1$$

Reflection occurring at the minimum value of θ is called *reflection of the first order;* that at the next larger angle — *reflection of the second order,* etc. The values 1, 2, 3, etc., correspond to the values of *n* in the equation.

$$n\lambda = 2d \sin \theta.$$

The Braggs built an instrument — an X-ray spectrometer — in accordance with the scheme outlined. Their work with this instrument resulted in the above formula and in many remarkable investigations of crystal structure.

3. Debye-Scherrer Method, or Powder Method. It has been said that in photographing by the Laue method polychromatic rays and a single crystal are used, because in the case of monochromatic rays there is very little chance of rays of a definite wavelength striking a plane net with a corresponding interplanar spacing at an angle θ required by the Bragg-Wulff formula, so that the reflected beam would be within the boundaries of the photographic plate.

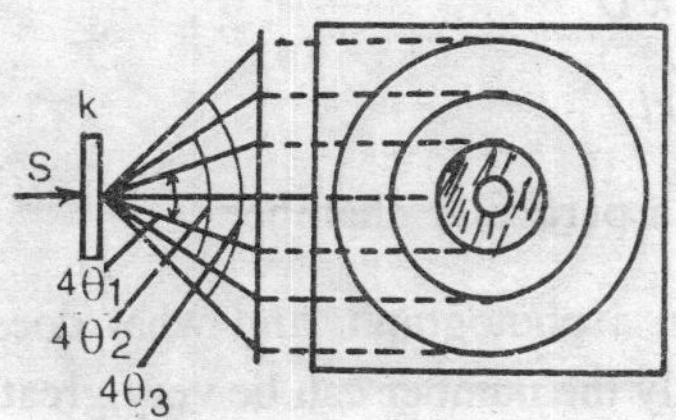

Fig. 6.4. Debye-Scherrer method: S—bundle of rays; *k*—crystal; M—central spot

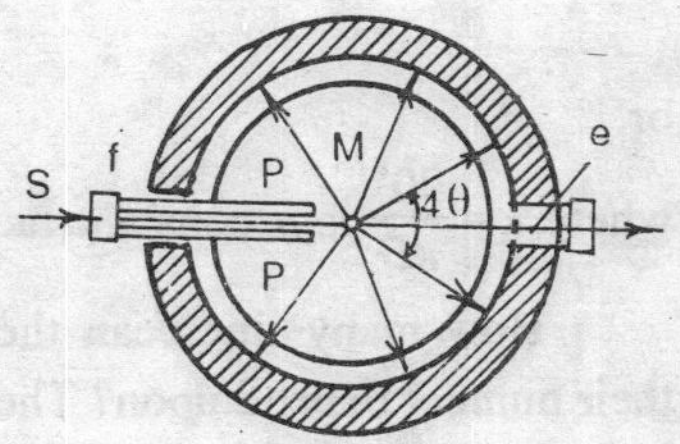

Fig. 6.5. Photographing by the Debye method.

Instead of a single crystal Debye and Scherrer began using powdered crystals, *i.e.*, a multitude of minute crystals, and monochromatic rays. Compressed into a rod sample, the tiny crystals have different orientations, and therefore the possibility of interference which depends on the ratio of λ, d and θ greatly increases. As a result, instead of individual spots whole series of spots are obtained (Fig. 6.4) which, merging, produce curves on a photographic plate or film.

The chamber is cylindrical in shape (Fig. 6.5), and the photographic film *PP* is bent along its inner surface. Primary rays enter the chamber through the diaphragm *f,* strike the sample *M* which is set up perpendicularly to the plane of the drawing, and leave through the aperture *e.* Secondary rays interfere diverging from the sample. The rays that intensify each other produce a number of cones interesting the cylindrically bent photographic film *PP*. At the intersection of the cones and the film, curves are obtained which are called *arcs* or simply *lines*. Figs. 6.4 and 6.5 show that the angles between the arcs of the same cone are 4θ.

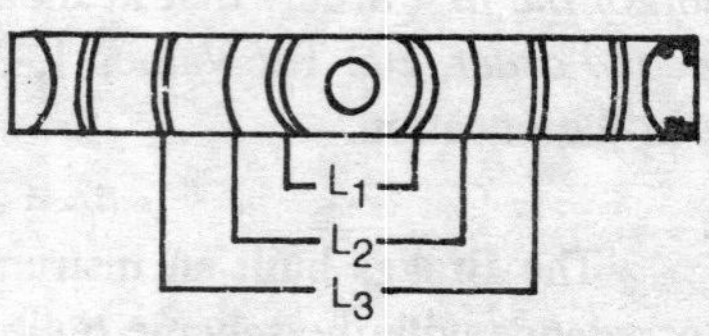

Fig. 6.6. A powder photograph (debyegraph).

On a stretched film the arcs appear as shown in Fig. 6.6. If the distances between symmetrical lines are designated as l_1, l_2, l_3, etc., and the diameter of the cylindrical film as *D*, the relation

$$\frac{l}{\pi D} = \frac{4\theta}{360^\circ}$$

will show that

$$\theta = \frac{90^\circ}{\pi D}\, l,$$

or

$$\theta = kl$$

where $k = \dfrac{90^\circ}{\pi D}$ is a constant factor for a particular chamber.

How many lines can there be in a photograph, and what does their number depend upon? Theoretically the number can be very great, as it depends on an infinitely large number of values of *d*, but actually it is relatively small for the following reasons.

The values of *d* are repeated in symmetrical lattices. For instance, in a primitive cubic lattice *d* is the same in all the three mutually perpendicular directions. Crystals oriented in the rod at a corresponding angle θ with respect to the primary beam will give similar beam cones in three positions, therefore the intensity of the arcs will be at its greatest. There are two such positions for tetragonal lattices. In orthorhombic lattices all the three principal interplanar spacings are different and the crystals give a greater number of lines, but of lower intensity. The same is true of the monoclinic and triclinic lattices. Hence, a small number of high-intensity lines denotes a cubic lattice. In the intermediate systems the number of lines increases and their intensity decreases. The lower systems are characterized by a very large number of low intensity lines.

The Debye-Scherrer method has the advantage of not requiring large single crystals, and almost any substance can be ground into powder. The study and measurement of the distances between the lines and of the line intensities help to solve many important problems, for instance:

(a) a crystalline substance can be distinguished from an amorphous one: the former produces lines on a photograph, the latter does not;

(b) if photographs are made before and after treatment (by heat, mechanically, or otherwise) of a crystalline substance, their comparison will reveal any alterations in the lattice;

(c) with the aid of standard photographs of known chemical compounds their presence may be detected in an unknown mixture;

(d) a substance, *e.g.*, a mineral, can be determined if its powder photograph is compared with the standard photographs in a mineralogical determinative manual;

(e) the size of the unit cell and the type of the lattice can be determined in simple cases.

4. Rotation Method is a combination of the Laue and Debye-Sherrer methods. The use of a single crystal identifies this method with the former, except that the crystal is definitely oriented; the use of monochromatic rays identifies it with the latter, this making it possible to use the Wulff-Bragg formula in interpreting photographs, since the wavelength λ is known.

This method requires a small crystal, preferably somewhat elongated, of 0.2-0.05 mm in cross-section. During exposure the crystal is rotated at a speed of 10-15 revolutions per hour. The axis of rotation should coincide with an edge of the crystal, *i.e.*, with a row of the lattice.

Plane nets of a crystal may be positioned at various glancing angles θ to the primary rays. At some values of θ interference occurs, intensifying the reflected rays, which produce a series of cone surfaces (Fig. 6.7). The latter, crossing a cylindrical photographic film P, produce a number of horizontal lines (Fig. 6.8) which are called *layer lines*.

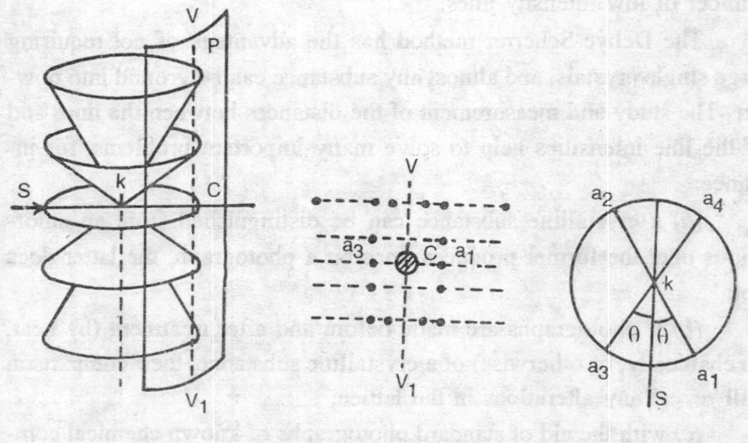

Fig. 6.7. Rotation method.

Fig. 6.8. X-ray photograph of rotation.

Fig. 6.9. Rotation method.

Each net gives two spot which are symmetrical with respect to the central vertical plane VV_1 (see Figs. 6.8 and 6.9 a_3 and a_1, because during a complete revolution of the crystal *k* the net twice occupies a position at the same angle θ to the incident rays *S* (Fig. 6.9).

Since in this case it is a single crystal that is photographed, and not powder as in the Debye-Scherrer method, the cone surfaces are intermittent and not continuous, and reflections are recorded on the film only at definite points. The location of these points corresponds to the plane net position required for an increase in intensity caused by interference at a definite value of the angle θ.

Nets which are parallel to the rotation axis give spots on the so-called *zero line* which passes through the central spot of the photograph. Nets which are inclined to the axis of rotation give lines located symmetrically above and below the zero line. Nets which are perpendicular to the axis of rotation give no reflections.

The VV_1 plane which is parallel to the axis of rotation and passes through the central spot *C* is called the principal plane (see Figs. 6.7 and 6.8).

If the direction of the axis of rotation is parallel to the dense row of the lattice and therefore the layer lines are perpendicular to it, the parameter *J* of this row, *i.e.*, its interplanar spacing, is determined from the formula:

$$J = \frac{n\lambda}{\sin \mu_n},$$

where *n* is the number of the layer line (counting from the zero line);

μ_n is an angle obtained from the equation.

$$\tan \mu_n = \frac{e_n}{R},$$

where e_n is the distance to the *n*th layer line from the zero line;

R is the radius of the cylindrical film.

If three photographs can be taken, the rotation method enables the size of the crystal cell to be easily determined. The crystal should be so oriented that in each case the axis of rotation if parallel to one of the three principal edges of the elementary parallelopiped.

7

Packing of Solid Spheres

7.1. Close Packing of Identical Solid Spheres

From the study of crystals it follows that a large number of ways are known in which identical solid spheres can be packed together. This close packing is essential due to the following two reasons:

(*i*) The spheres prefer closest packing possible so that a maximum possible density is reached.

(*ii*) They prefer closest packing on account of their mutual attractive interactions because closer they lie, the greater is the stability of the packed system.

We know that the constituent particles are having various shapes. Therefore, there will be different types of packing according to their shapes.

Let us now discuss the different packing modes of simple spherical particles of almost equal size.

First of all, the packing of a number of spheres in one plane is considered. This arrangement is depicted in Fig. 7.1. The arrangement shown in Fig 7.1 (A) is more economical than that shown in Fig. 7.1 (B). This is the closest packing and hence is the most stable arrangement possible. From calculations, it is evident that in this arrangement only 60.4 per cent of the space has been occupied by the spheres while the remaining 39.6 per cent of the space is empty and is termed as *void volume*. As any of the spheres is having six nearest neighbours in this arrangement, the coordination number of each sphere when packed in this manner is six. Thus, a sort of a hexagonal pattern is formed around each sphere which is depicted in Fig. 7.1 (C) and (D).

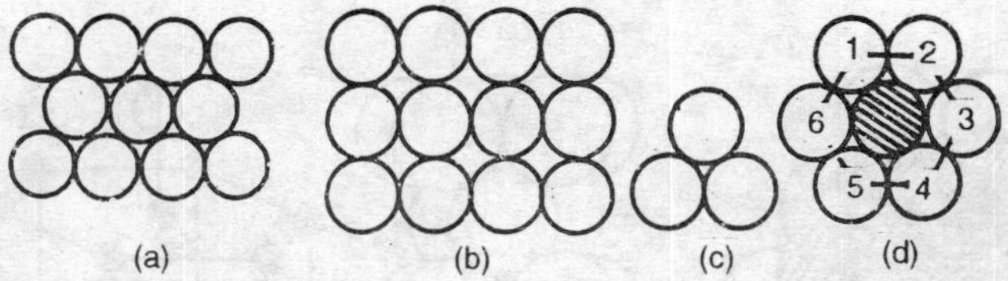

Fig. 7.1. Close packing of spheres in one plane.

Let us now construct another layer on the first lower layer. One method is to fit a sphere above a hollow at the centre of three touching

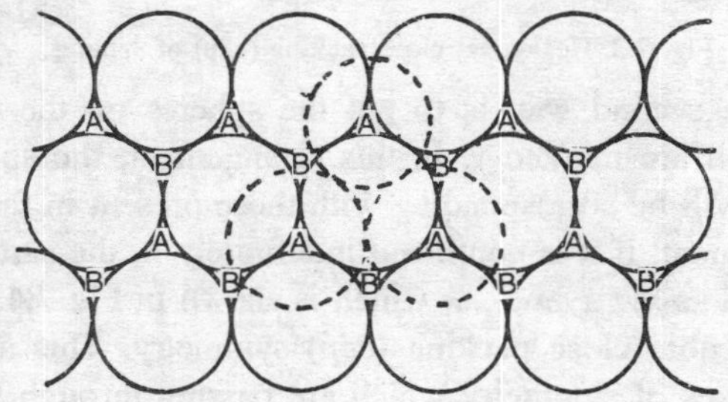

Fig. 7.2. Close packing of spheres in two layers.

spheres in the first layer. Thus, four spheres, three first layer and one of second layer, are available. The centres of these spheres will occupy the four corners of a regular tetrahedron. This type of arrangement is depicted in Fig. 7.2 in which the full-line circles denote the spheres in the first layer while the dotted circles denote those in the second layer. In each layer, there occurs closest possible packing.

In order to build up the third layer, there are two different ways to do so. These are as follows:

(*i*) One way to do so is to place the spheres vertically above those in the first layer. Thus, the spheres of the third layer will be present on one set of hollows marked '*x*'. If this arrangement is continued indefinitely in the same sequence, this is represented as *xy xy xy x*. This type of arrangement reveals that every third layer of the spheres is situated directly over the first layer.

When we examine this arrangement carefully, it is found to represent hexagonal close packing (*hcp*) symmetry which means that the whole structure has only one-six fold axis of symmetry. The same appearance has also been realised when the crystal is rotated through an angle of 60° (Fig. 7.3). It is to be mentioned here that the symmetry axis is perpendicular to the layers of close packed spheres.

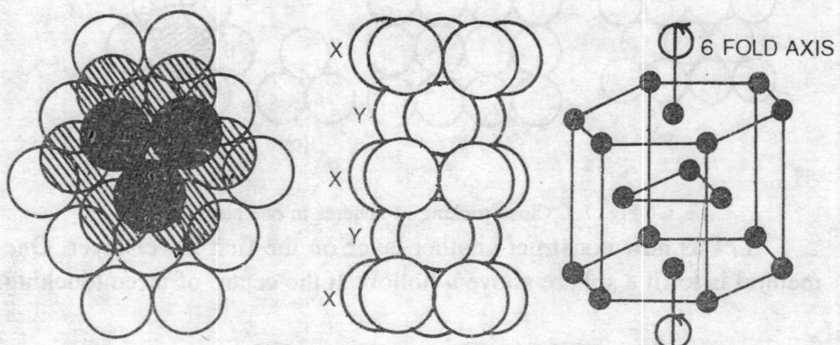

Fig. 7.3. Hexagonal close packing (hcp) of spheres.

(*ii*) The second way is to put the spheres on the other set of hollows which are marked *y*. In this arrangement, the spheres in the fourth layer will be corresponding with those present in the first layer. This arrangement, if it is continued indefinitely in the same sequence, is represented as *xyz xyzx*......... which is shown in Fig 7.4. This arrangement has cubic close packing (*ccp*) symmetry. This structure has four 3-fold axes of symmetry which are passing through the diagonal of the cube. One of these axes is shown in Fig. 7.4.

On examining the *xyz xyzx* .... system of close packing carefully, it is observed that there is a sphere at the centre of each face of the unit cube and therefore this structure is also called face centred cubic structure (Fig 7.4). In this arrangement, each sphere has been found to be in contact with the six nearest spheres. Also, each sphere is also found to be in contact with three spheres in layer above and three spheres in the layer below. Hence, in both *hcp* and *ccp* arrangements each sphere has been found to be surrounded by twelve equidistant spheres. Hence, the coordination number in each case is 12.

By calculations, it has been shown that only 74 per cent of the

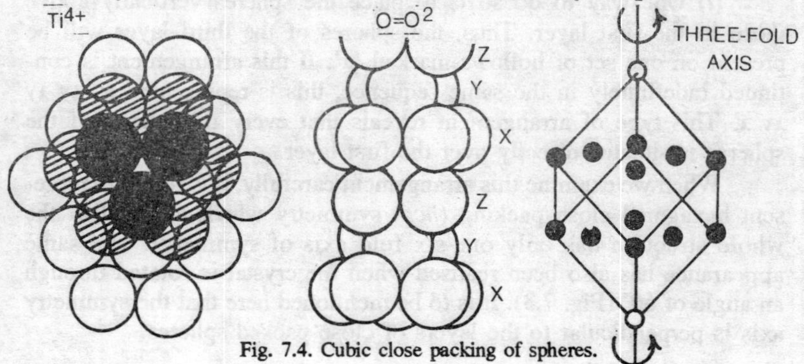

Fig. 7.4. Cubic close packing of spheres.

total space has been occupied by the spheres while the remaining 26 per cent of the space is empty and is known as **void volume.**

There is another possible arrangement of packing of spheres called body-centred cubic (*bcc*) arrangement which will be only formed when the spheres in the first layer (marked *x*) of cubic closest packing undergo slight opening. Due to this, none of these will remain in contact with each other. Such an arrangement is depicted in Fig 7.5(*a*). The second layer of the spheres (marked *y*) may be built up on top of the first layer in such a way that each sphere of the second layer has been found to be in contact which four spheres of the layer below it. If the successive building of the other layers is carried on, third layer will be same as the first layer, *i.e.*, on the top of *x*. If this pattern of building layers is repeated infinitely, an arrangement shown in Fig. 7.5(*b*) would be obtained. In this arrangement, one sphere is present at each corner and also at the centre of each unit cube. Hence, in this arrangement, each sphere has been found to be in contact with eight other spheres (four spheres in the layer just above and four spheres in the layer just

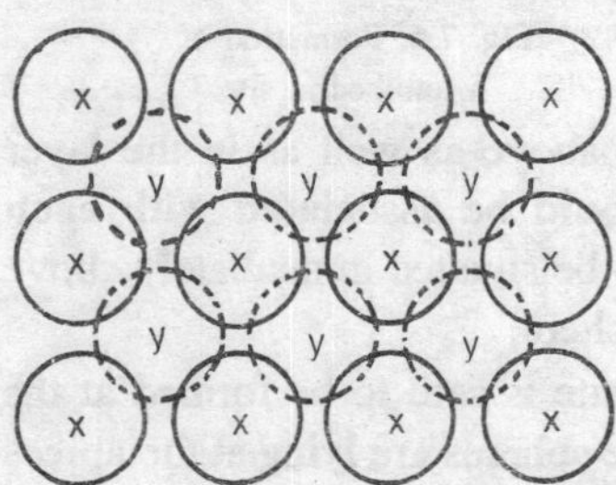

Fig. 7.5. (a) Body centred cubic packing of spheres.

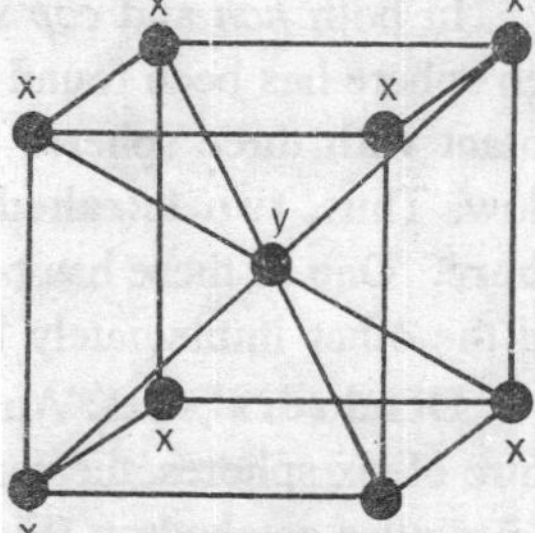

Fig. 7.5 (b) *b.c.c.*

below) so that the coordination number in this type of arrangement would be eight.

Most of the metallic elements have crystal lattices belonging to one or the other of the above three types. Some examples of these with their coordination numbers have been given below:

Metallic Elements	*Coordination Number*	*Lattice Type*
Be, Mg, Cd, Zn, Ti	12	Hexagonal close packing (*hcp*)
Cu, Ca, Sr, Ag, Au	12	Cubic close packing (*ccp*)
Li, Na, K, Rb, Ca	8	Body centred cubic arrangement (bcc).

Interstitial sites in crystals. Before we describe the structure of ionic compounds further, it would be appropriate to understand for the formation of interstitial sites in between the closely packed spherical structures.

Tetrahedral sites. A tetrahedral structure is said to be formed if one sphere is kept upon three other spheres which are touching one another (Fig. 7.6). But the four spheres are touching each other at one point only. Therefore, a small space called *tetrahedral site* is left in between these spheres. The size of site is much smaller than that of the spheres. Further, the larger the spheres, the larger would be the size of the tetrahedral site formed by them.

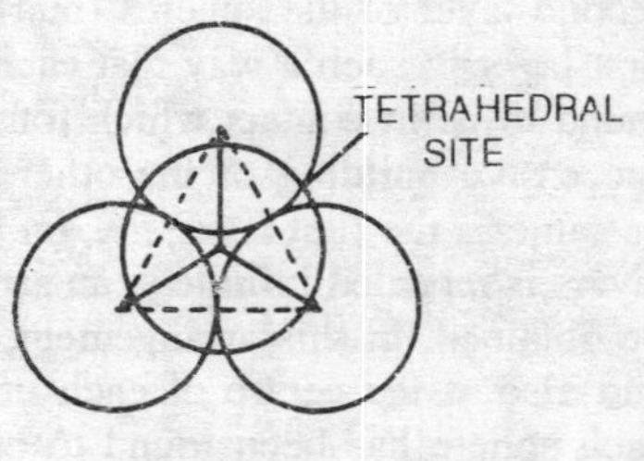

Fig. 7.6. Formation of a tetrahedral site.

In both *hch* and *ccp* systems, each sphere has been found to be in contact with three spheres in the layer above as well as in the layer below. Thus, **two tetrahedral sites would be associated with each sphere.** One of these has been found to be situated immediately above and the other immediately below the sphere.

Octahedral sites. An octahedral site is said to be formed at the centre of six spheres, the centres of these spheres are lying at the apices of a regular octahedron (Fig. 7.7). This site is formed in both *hcp* and *ccp* systems.

In Fig. 7.7, two layers of closepacked spheres all shown. The spheres in one place are represented by the full circles while the spheres in second plane by dotted circles. Two triangles are drawn; one is joining the centres of three spheres in one plane while other (dotted) is joining two centres three spheres in second plane. Wherever two triangles with apices in opposite direction of different layers get superimposed one above the other, the octahedral sites are said to be formed. These have been shown by '*A*'.

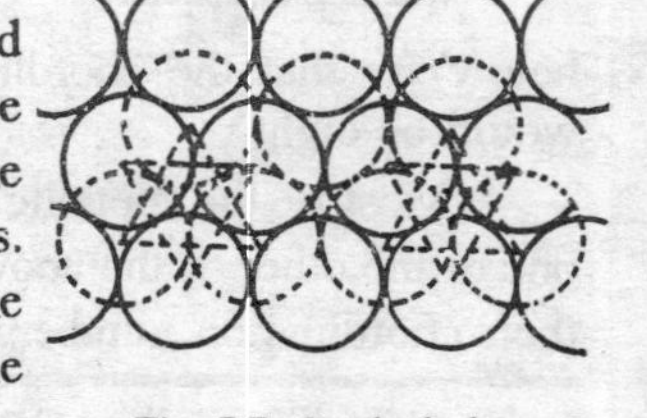
Fig. 7.7. Octahedral interstitial sites.

In order to derive the relation between the size of an octahedral site and the size of the spheres surrounding it, a section of an octahedral

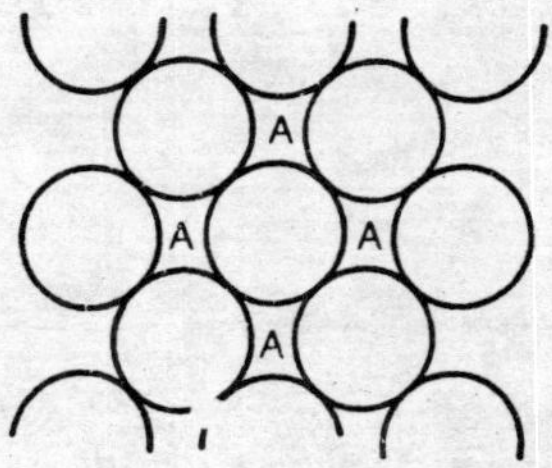

Fig. 7.8. Cross-section of octahedral interstitial site in a *fcc* lattice.

site surrounded by four spheres (shown in Fig. 7.9) is taken into the consideration. Suppose this site is occupied by a small sphere which is assumed to be touching the four surrounding spheres.

The centre X of the smallar sphere is joined with Y and Z which are centres of the two larger spheres. Now L represents the point of contact of the two larger spheres while XL is perpendicular to YZ.

Suppose r_1 is the radius of the smaller sphere while r_2 is the radius of the bigger sphere. From Fig. 7.9, it is evident that

Fig. 7.9. Cross-section through an octahedral site.

$$XL = YL + r_2,$$

But XYL is a right-angled triangle.
Therefore,

$$XY^2 = XL^2 + YL^2$$

$$(r_1 + r_2)^2 = r_2^2 + r_2^2$$

or $$(r_1 + r_2)^2 = \sqrt{2} r_2 = 1.414 r_2$$

or $$r_1 = 0.414 r_2$$

or $$r_1 / r_2 = 0.414$$

From the above relation, it is evident that the size of an octahedral size is only about 0.414 of the size of the surrounding spheres. Similarly, it may be proved that the size of a tetrahedral site is about 0.225 of the size of the surrounding spheres. Thus, *an octahedral site is larger than a tetrahedral site.*

8

Ionic Crystals and other Contents

8.1. Introduction

The forces which hold the ions together within the crystal are the electrostatic coulombic forces. The type of lattice with which the ionic compound crystallises depends upon two factors:

(*a*) The size of the ions, and

(*b*) The necessity for the preservation of electrical neutrality, different sizes. However, the two types of charges are present in equivalent amounts. The voids in ionic crystals will be much wider than those in packing of equal spheres as in metals. In an ionic crystal, there are two inter-penetrating lattices of anions and cations.

A typical example of ionic crystals is sodium chloride, NaCl. From the X-ray study of this crystal, it is followed that sodium chloride has octahedral structure. Each Na^+ ion represented by a plane circle is surrounded by six Cl^- ions represented by shaded circles at the corners of a regular octahedron and similarly each Cl^- ion is surrounded by six Na^+ ions. Thus, the coordination number of Na^+ ion as well as that of Cl^- ion, is 6.

In NaCl, it has been determined that the distance between the adjacent ions is 2.81 Å. If we add the radii of sodium (0.95 Å) and chlorine (1.81 Å), it shows that there is a little space between the ions.

All alkali halides, except those of cesium, crystallise in the form of sodium chloride structure.

As ionic crystals possess different types of constituent particles in neighbouring positions, this obstructs any easy slip between neighbouring units in ionic crystals and, therefore ionic crystals are liable to break under stress. In an ionic crystal lattice strong electrostatic forces hold constituent ions together. Therefore, ionic solids possess fairly high melting points.

8.2. Coordination Number

In an ionic solid, positive ions are surrounded by negative ions and vice versa. *The number of ions of opposite charge surrounding an ion in the crystal lattice is the coordination number of that ion.* For example :

(*i*) In sodium chloride, each Na^+ ion is surrounded by six Cl^- ions and similarly each Cl^- ion is surrounded by six Na^+ ions. Thus, the coordination number of both Na^+ and Cl^- ions is six.

(*ii*) In calcium fluoride, each Ca^{2+} ion is surrounded by eight F^- ions but each F^- ion is surrounded by four Ca^{2+} ions. Thus, the coordination number of Ca^{2+} is eight whereas of F^- is four.

From the above examples, it follows that the coordination numbers of A^+ and B^- must be equal in a binary compound, AB (NaCl, KCl) but they are not equal in AB_2 compounds (CaF_2, TiO_2) in which the coordination number of A^{2+} will be twice that of B^-.

Influence of Coordination Number. The radius of a given ion is influenced by the coordination number. Thus, for instance, the radius of Cl^- ion in sodium chloride [coordination number of Na^+ and Cl^- is six] is somewhat less than in cesium chloride [coordination number of Cs^+ and Cl^- is eight]. This may be explained by considering the forces of attraction and repulsion between different ions ; in cesium chloride, each ion is surrounded by a greater number of ions as compared with sodium chloride and this results in the less attraction to any one of them, and consequently, the greater will be the effective ionic radii.

The value of relative ionic radius increases with coordination number. Numerically, this fact can be represented as follows:

Co-ordination number	4	6	8
Radius (relative)	9.95	1.0	1.0

Generally, the co-ordination number of sodium chloride structure, *i.e.*, six, is taken as standard and the values of ionic radii are generally expressed with respect to this co-ordination number.

8.3. Radius Ratio

The co-ordination number and geometry of an ionic crystal depend upon the radius ratio *which is defined as the ratio of radius of cation to that of anion.* This can be put as

$$R_r = \frac{r_{c^+}}{r_{a^-}}$$

where R_r is called the radius ratio, and r_{c^+} and r_{a^-} are the ionic radii of the cation and anion of the ionic crystal $c^+ a^-$ respectively.

The effect of radius ratio in determining the co-ordination number and shape of an ionic crystal is known as **radius ratio effect.**

In order to understand the concept of radius ratio, an ionic compound AX is considered. The coordination number of A^+ ion in this compound is *three, i.e.*, one A^+ ion is surrounded by three X-ions. This has been shown in Fig. 8.1(a). This structure is very stable due to the following reasons:

(*i*) As the positive ion (A^+) is an contact with three negative ions (X^-), the forces of attraction will be strong.

(*ii*) As the negative ions are not at all touching one another, the forces of repulsion will be quite small.

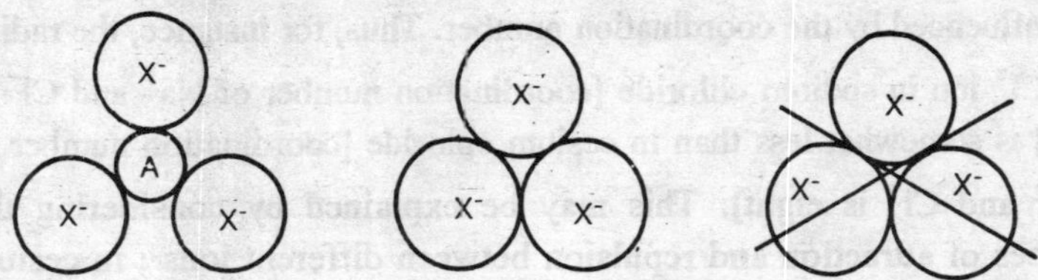

Fig. 8.1. Size of A+ and the radius ratios.

Suppose the negative ions (X^- ions) are in contact with one another [Fig. 8.1 (b)]. This is the limiting case of stability with coordination number 3. In this case, the positive ion is still touching the negative ion. Therefore, the force of attraction is quite strong. But at the same time the negative ions are also touching one another. Therefore, the force of repulsion will be stronger than before. The radius ratio for such a case has been found to be 0.155. Suppose the A^+ ion still becomes smaller. In that case the radius ratio falls below this value (*i.e.*, 0.155) and the structure will be as shown in Fig. 8.1 (c). This structure is unstable due to the following reasons :

(*i*) In this structure, two negative ions are in direct contact with one another. Therefore, the force of repulsion will be quite strong.

(*ii*) Also, the positive ions is not in direct contact with negative ions. Therefore, the force of attraction will be quite weak.

Due to strong repulsive force, the structure shown in Fig. 8.1 (c) is not stable.

It also follows from the above discussion that if the value of radius ratio falls below 0.155 for coordination number 3, the structure will be unstable. Thus, the limiting radius ratio is 0.155.

For each co-ordination number and geometry of the ionic crystal the value of radius ratio is calculated if the ionic radii are known. This radius ratio is known as the *limiting radius ratio.* If the radius ratio of an ionic crystal falls below the limiting radius ratio, then the structure is unstable.

Co-ordination numbers of 1, 4, 6 and 8 are common, the appropriate limiting radius ratio can be worked out and their corresponding shapes may be predicted. This information is shown in the following table.

Table 8.1

Co-ordination number	Shape	*Limiting radius ratio*
2	Linear	0 to 0.155
3	Plane triangle	0.155 to 0.225
4	Tetrahedral	0.225 to 0.414
4	Square planar	0.414 to 0.732
6	Octahedral	0.414 to 0.732
5	Body-centred cubic	0.732 to 1.000

Let us apply the concept of radius ratio to the different examples.

(*i*) The radius ratio of sodium chloride is 0.52. It appears from table, that it should have octahedral arrangement. Experimentally, it has been found to be so.

(*ii*) In zinc sulphide, the limiting radius ratio is 0.40. This suggests a tetrahedral arrangement in which each Zn^{2+} ion is tetrahedrally surrounded by four S^{2-} ions and each S^{2-} is tetrahedrally surrounded by four Zn^{2+} ions. Actually, this has been found to be so.

(*iii*) In cesium chloride, the radius radio is 0.98, indicating a body centred cubic arrangement where each Cs^+ is surrounded by eight Cl^- ions and vice versa. This has been confirmed experimentally.

We will now discuss the structures of few ionic solids of the type AX and AX_2.

8.4. Structure of Ionic Solids of General Formula AX

Examples of such ionic solids are sodium chloride, cesium chloride and zinc sulphide. These are described below:

Sodium chloride structure. As the radius of Na^+ ion is 0.95 and that of Cl^- ion is 1.81 Å, the radius ratio 0.95/1.81 or 0.524. This value lies between and 0.732, 0.414 thereby revealing that the co-ordination number should be 6 or 4 and the corresponding shape either **octahedral or square planar.**

When X-ray study of sodium chloride crystal was carried out it was found that this has octahedral structure. Each Na^+ ion is surrounded by six Cl^- ions at the corners of a regular octahedron and also each Cl^- ion is surrounded by six Na^+ ions. Thus, the co-ordination number of both Na^+ and Cl^- ions is six. This arrangement is shown in fig. 8.2.

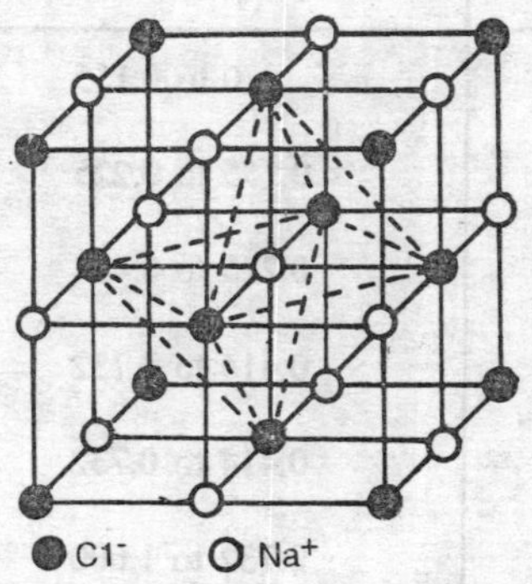

Fig. 8.2. Octahedral structure of NaCl crystal.

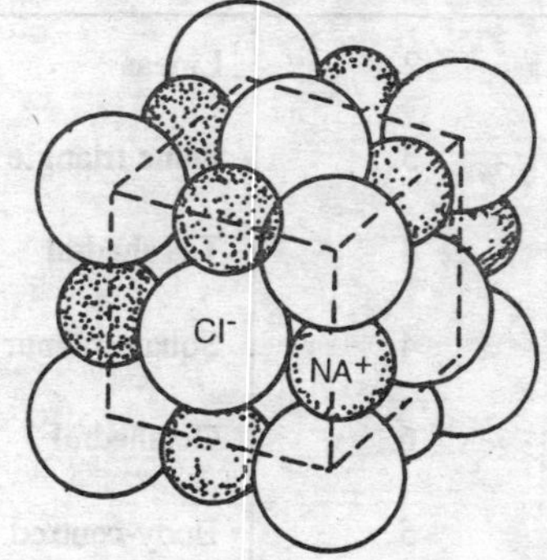

Fig 8.3. Actual space filling model of NaCl crystal.

From Fig. 8.2, it appears that sodium and chloride ions are not in direct contact with each other. Actually, these ions are in direct contact with each other as shown in Fig. 8.3 which is confirmed by

X-ray study. This study also shows that the distance between the centres of Na+ and Cl^- is 2.814 Å. But the radii of Na+ and Cl^- ions are 0.95 and 1.81 Å. Therefore, the result is 0.95+1.81 or 2.76 Å. This value is very close to 2.814 Å, thereby confirming that there is little or no space between the adjacent ions.

In this structure, chloride ions are in *ccp* type of arrangement, *i.e.*, it is having chloride ions at each face of the cube. On the other hand, sodium ions are so located that each Na^+ ion is surrounded by six chloride ions. This is equivalent to saying that sodium ions are occupying all the octahedral sites (holes constituent by the packing of Cl^- ions). As there is only on octahedral site for every chloride ion, the stoichiometry is 1 : 1. This required from the stoichiometry of Na^+ Cl^- as well.

It is important to mention here that neither Na^+ ions are touching each other nor Cl^- ions are touching each other. Had they been touching each other, the structure would have been stable. The Na^+ ion present in octahedral sites are quite large to separate the Cl^- ions from one another. Similarly, the Cl^- ions are sufficiently large to separate Na^+ ions from one another.

Except cesium chloride, all other alkali halides are having sodium chloride structure. Also, alkaline earth metal oxides and sulphides (except ZnS), selenides and tellurides (except magnesium selenide) are having sodium chloride structure.

Cesium Chloride Structure. The radius ratio of cesium chloride is 1.69/1.81 or 0.93. This value suggest that it has a coordination number of 8 and 9 cubic structure. The Cl^- ions form the simple cubic arrangement while cesium ions occupy the cubic interstitial sites, *i.e.*, each cesium ion is having eight chloride ions as its nearest neighbours. From Fig. 8.4, it can be observed that each chloride ion is also surrounded by eight cesium ions which are also disposing towards the corners of a cube. Thus, both types of ions in cesium chloride are in equivalent positions and the stoichiometry is 1 : 1 and the coordination number is 8 : 8.

As the coordination number of CsCl is higher than that of NaCl, it is expected that cesium chloride is more stable that sodium chloride because each cesium ion is having more ions of opposite charge as its neighbours. Actually, cesium chloride lattice has been found to be one per cent more stable than sodium chloride lattice.

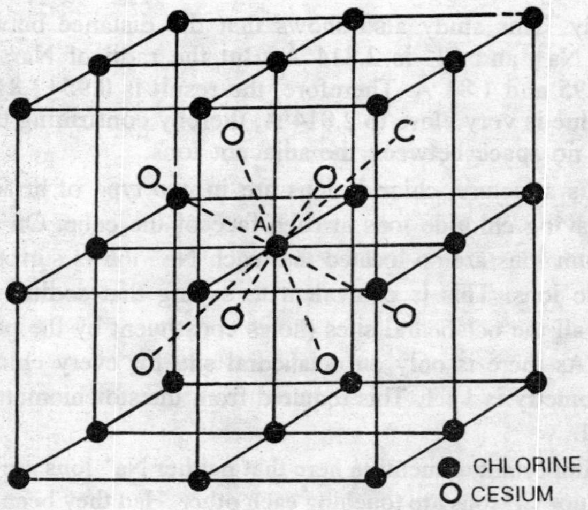

Fig. 8.4. Body centred cubic structure of CsCl.

Bromides and iodides of cesium are having the same coordination number and the same structure as that of cesium chloride.

Zinc Sulphide Structure. It is example of AX type of ionic solids. It exhibits two types of structures such as (i) zinc blende structure and (ii) wurtzite structure.

(i) *Zinc blende structure*. As the radius ratio of zinc sulphide is 0.40, this reveals that the coordination number is 4 and it is having a tetrahedral structure. Hence, each Zn^{2+} ion is being surrounded by four

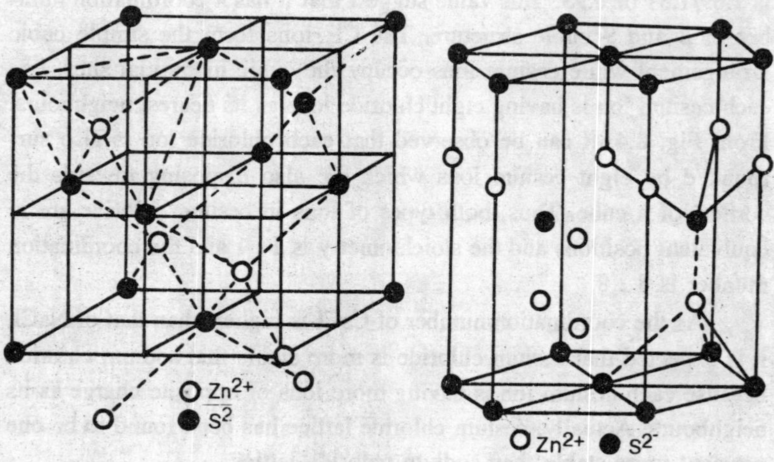

Fig. 8.5. (a) zinc blende structure. (b) Wurtzite structure.

S^{2-} ions which are disposed towards the four corners of a regular octahedron. Also, each S^{2-} is being surrounded by Four Zn^{2+} ions [Fig. 8.5 (a)]. The sulphur atoms are in *ccp* (cubic close packed) type of arrangement and zinc atoms are occupying octahedral sites.

As there are two tetrahedral sites available for every sulphur atoms, zinc atoms are occupying half of the tetrahedral sites while the remaining tetrahedral sites are empty. Thus, only alternate tetrahedral sites are filled by zinc atoms and the stoichiometry of the compound is 1 : 1.

(ii) *Wurtzite structure.* This structure is shown in Fig. 8.5 (b). In this structure, sulphide ions are forming the *hcp* type of arrangement. In this type of arrangement, the two tetrahedral sites are available for every sulphide ion and, therefore, zinc ions are occupying only half of the tetrahedral sites. The alternate tetrahedral sites remain vacant. The stoichiometry of the compound is 1 : 1.

The main difference between zinc blende and wurtzite structures is that the former is having *ccp* type of packing for sulphide ions while the latter is having *hcp* type of packing for sulphide ions. However, the coordination number of Zn^{2+} and S^{2-} in either case is 4 : 4.

Other examples of Wurtzite type of structure are ZnO, CdS, CdSe and BeO.

8.5. Structure of Ionic Solids of General Formula, AX_2

Examples of these are calcium fluoride, titanium dioxide, calcium carbide and titanium dioxide. We shall discuss these one by one.

Calcium Fluoride structure. This structure is shown in Fig. 8.6. The fluorite structure is found when the radius ratio is 0.73 or above.

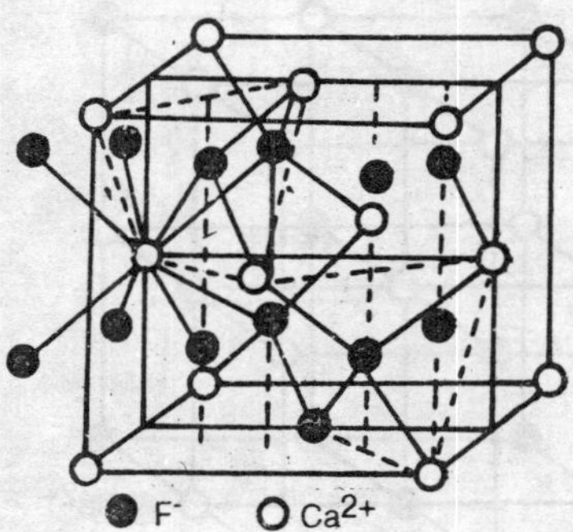

Fig. 8.6. Body-centred cubic structure of calcium fluoride structure.

In the fluorite structure, each Ca^{2+} ion is surrounded by eight F^- ions at the corners of a body centred cubic arrangement while each F^- ion is surrounded by four Ca^{2+} ions arranged tetrahedrally. Thus, the coordination numbers of Ca^{2+} and F^- are 8 and 4. Such an arrangement is known 8 : 4 arrangement. Calcium ions in fluorite structure are arranged in face-centred cubic close packed (*ccp*) type of arrangement. As there are two tetrahedral sites available for every calcium ion, the fluoride ions occupy all the tetrahedral sites. Thus, the stoichiometry of the compound is 1 : 2.

Others compounds showing this type of structure are SrF_2, BaF_2, $BaCl_2$, $SrCl_2$ CdF_2, HgF_2 and PbF_2.

Structure of titanium dioxide, TiO_2 (Rutile structure)

The rutile structure is shown in Fig. 8.7. In this structure, each Ti^{2+} ion is surrounded by six O^{2-} ions arranged octahedrally and each O^{2-} ion is surrounded by three Ti^{4+} ions arranged in plane triangular manner. Thus, the coordination numbers of Ti^{4+} and O^{2-} are 6 and 3.

The structure of titanium dioxide cannot be considered to cubic as one of the axes is shorter than the other by 30 per cent.

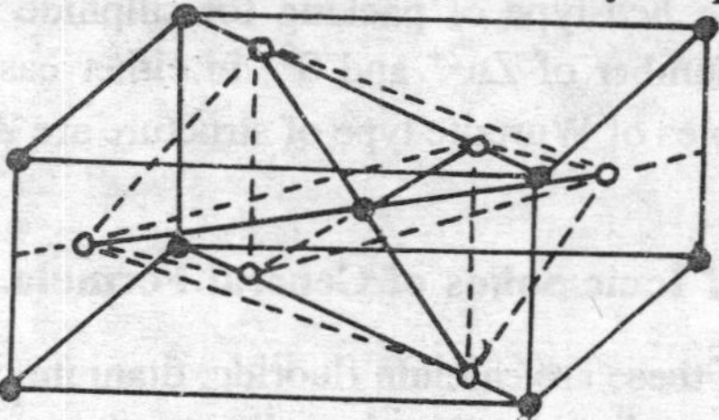

Fig 8.7. Structure of titanium dioxide (Rutile).

Structure of Calcium Carbide. This structure is shown in Fig. 8.8. This structure is similar to that of sodium chloride in

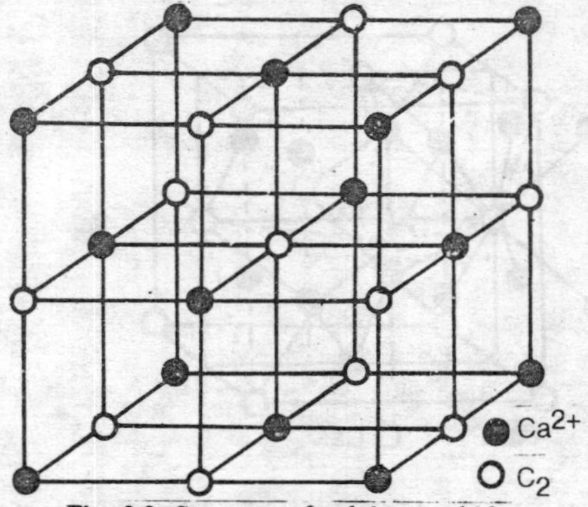

Fig. 8.8. Structure of calcium carbide.

which sodium ions are replaced by calcium ions while chlorine ions by carbide ions. The carbon atoms in carbide are associated in pairs which are aligned in parallels.

Unlike sodium chloride, the cubic symmetry is distorted in calcium carbide.

The structure of FeS_2 has been found to be similar to that CaC_2. However, Unlike CaC_2, the S_2 units in FeS_2 do not align in parallels.

Structure of Cadmium Iodide. This structure crystallises in so-called layer structure shown in Fig. 8.9. In this structure each cadmium atom is surrounded octahedrally by six iodine atoms while three cadmium atoms nearest to each iodine atoms are at the corners of a pyramid of which the iodine atoms is the apex. The atoms thus form layers, and each sandwich consisting of a layer of cadmium atoms with layers on either side of it, is electrically neutral. As there are only van der Waal's forces between adjacent compositae layers, crystals of this kind show a pronounced cleavage parallel to the layers. The unsymmetrical environment of the iodine atoms reveals that this structure is not a purely ionic one.

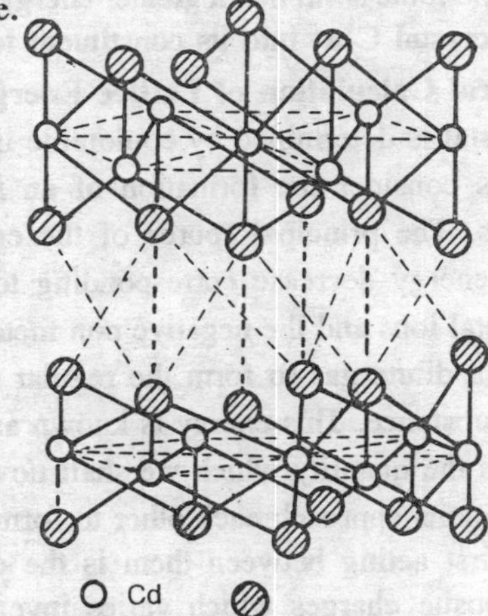

Fig. 8.9. Structure of cadmium iodide.

8.6. Lattice Energy

The lattice energy of an ionic solid is defined as *the energy released when the correct number of gaseous cations and gaseous anions are brought together from infinite distance to form one mole of an ionic solid.*

The formation of a mole of ionic solid $C^+ a^-$ from the constituent gaseous ions may be represented as

$$C^+ \text{ (g)} + a^- \text{ (g)} \rightarrow C^+ a^- \text{ (s)} + \text{Energy released.}$$

The energy released in the above equation is known as lattice energy. It is usually denoted by U_{ca}. According to the conventions of the 'First Law of Thermodynamics' the energy released *i.e.*, the lattice energy, has a negative sign.

It follows from the First Law of Thermodynamics that energy released in the formation of one mole of an ionic solid from the constituent gaseous ions is numerically equal to the energy that will be needed to convert one mole of the same ionic solid into the gaseous ions. Thus, the lattice energy may also be defined *as the energy required to remove ions of one gram mole of an ionic solid from their equilibrium position in the crystal to infinity.* This may be represented as

$$C^+a^- \text{ (Solid)} + \text{Energy required} \rightarrow C^+ \text{ (g)} + a^- \text{ (g)}$$

From the above definition, one can safely draw the important conclusion that the stability of an ionic solid is measured in terms of its lattice energy. It means that *greater the value of lattice energy, more stable is the ionic solid i.e.*, a greater energy is required to decompose the ionic crystal C^+a^- into its constituent ions, C^+ and a^-.

Theoretical Calculation of Lattice Energy. The lattice energy of an ionic crystal is determined by coulombic interaction between all its ions. Let us consider the formation of an ionic crystal from its constituent ions. The principal source of the energy evolved in this process is the energy decrease corresponding to the condensation of the possible metal ions and the negative non metal ions, widely spaced in the form of a dilute gas to form the regular crystal lattice characteristic of the substance. This energy is known as the *lattice energy* of the crystal; it is the quantity which we shall now examine in detail.

When the ions approach each other to form the ionic crystal, the only force at first acting between them is the electrostatic attraction due to the opposite charges which varies inversely with distance, r between the two ions.

$$\text{P.E.}_{\text{(att)}} = \frac{Z_1 Z_2 (+e)(-e) A}{r} = -\frac{Z_1 Z_2 e^2 A}{r} \quad ...(1)$$

where Z_1 and Z_2 are the number of charges on cations ($+e$) and anions ($-e$) respectively. A is known as **Madelung** constant which depends on the geometric arrangement of the ions in the lattice. Since the value of this constant depends only on the geometry of the crystal, the value will be the same for all other solids which exhibit the sodium chloride structure or some other structure.

As the ions approach close to one another however, this attractive force given by equation (1) is opposed by the repulsive force due to

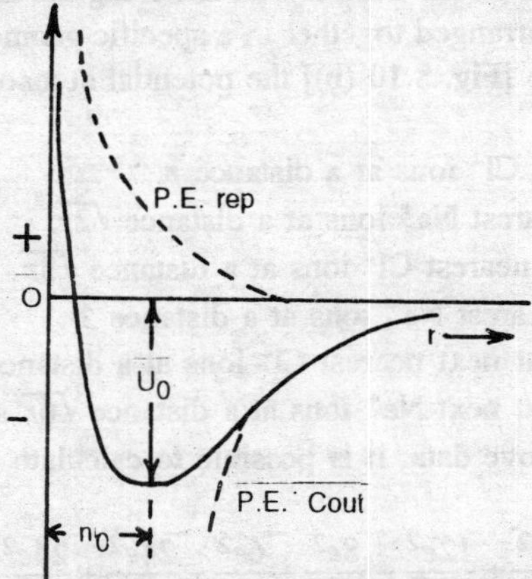

Fig. 8.10 (a) Plots of the different energy terms against *r* (qualitative only) giving a minimum at the internuclear distance r_0.

the electron systems of the two ions which is given by equation (2) [Fig. 8.10 (a)].

$$\text{P.E.}_{\text{(repulsion)}} = \frac{Be^2}{r^n} \qquad ...(2)$$

where *e* is the electronic charge, *n* is called the *Born exponent* and *B* is a *repulsion coefficient* which measures the strength of the repulsive force and depends upon the particular ion present. The value of *n* depends upon the types of electronic configuration which are present in the ions of the crystal; *n* generally as a valuc between 9 and 12.

When the ions are brought together from infinite distance, the net potential energy for ions is obtained by adding Eqs. (1) and (2).

$$\text{P.E.} = -\frac{Z_1 Z_2 e^2 A}{r} + \frac{Be^2}{r^n} \qquad ...(3)$$

Equation (3) is known as Born equation. This equation gives the potential energy for a cation carrying Z_1 as integral charge and potential energy for an anion carrying Z_2 as integral charge. From Eq. (3) it is also evident that the repulsion term, (be^2/r^n), increases more rapidly than the first term with decrease in the value of *r*. Born equation (3) is mainly used for calculating the *energy released* if a cation and an anion, which are separated by an infinite distance in gaseous state, are put together in a crystal at a distance *r* from each other.

We will now illustrate the application of Born equation to a sodium chloride structure. This crystal is having too many cations and anions which are arranged together in a specific geometry. In a crystal of sodium chloride [Fig. 8.10 (b)] the potential at a sodium ion results from

(i) Six nearest Cl^- ions at a distance r,

(ii) Twelve nearest Na^+ ions at a distance $\sqrt{2r}$.

(iii) Eight next nearest Cl^- ions at a distance $\sqrt{3r}$.

(iv) Six next nearest Na^+ ions at a distance $2r$.

(v) Twenty four next nearest Cl^- ions at a distance $\sqrt{5r}$.

(vi) Twenty four next Na^+ ions at a distance $\sqrt{6r}$ and so on.

From the above data, it is possible to calculate potential energy by using Eq. (3).

$$(\text{P.E.})_1 = \frac{6e^2}{r} + \frac{12e^2}{\sqrt{2r}} - \frac{8e^2}{\sqrt{3r}} + \frac{6e^2}{2r} - \frac{24e^2}{\sqrt{5r}} + \frac{24e^2}{\sqrt{6r}} \cdots$$

$$= -\frac{e^2}{r}\left[6 - \frac{12}{\sqrt{2}} + \frac{8}{\sqrt{3}} - \frac{6}{2} + \frac{24}{\sqrt{5}} - \frac{24}{\sqrt{6}} \cdots\right] \quad \text{...(3A)}$$

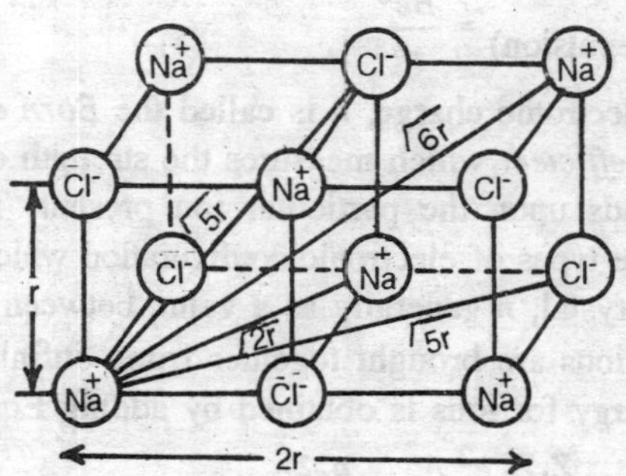

Fig. 8.10(b). Interionic distances in NaCl structure.

In the above equation, the product Z_1Z_2=1 because Z_1 and Z_2 for a uni-univalent electrolyte, *i.e.*, NaCl, are equal to unity; the potential energy due to repulsion is given a + ve sign.

The quantity in bracket of Eq. (3A) is the summation of an infinite series and is called *Madelundg constant* for sodium chloride structure. The summation value for sodium chloride comes out to be 1.747558. This value is the Madelung constant for sodium chloride. As the value of this constant depends only on the geometry of the crystal, it means that all other salts having sodium chloride geometry will have the same value of Madelung constant.

The Madelung constant is represented by symbol A. Its value for other salts can be calculated provided the geometrical arrangement of the ions in these salts is known.

The values of *A* for some common crystal structural types are given in the following table :

Table 8.2

Structure	*Formula*	*Coordination Number*	*A*
Sodium chloride	NaCl	Na : 6; Cl : 6	1.7475
Cesium chloride	CsCl	Cs : 8; Cl : 8	1.7627
Zinz blende	ZnS	Zn : 4; S : 4	1.6381
Wurtzite	ZnS	Zn : 4; S : 4	1.6410
Rutile	TiO_2	Ti : 6; O : 3	4.8160
Cadmium iodide	CdI_2	Cd : 6; I : 3	4.7100

Let us now again come to Eq. (3).

At the equilibrium distance, r_0, the attractive and repulsive forces are exactly balanced and the potential energy becomes minimum at r_0. In order to calculate this, differential equation (3) with respect to *r* is then put $\frac{d\,(\text{P.E.})}{dr} = 0$ for $r = r_0$

$$\frac{d\,(\text{P.E.})}{dr} = -\ Z_1 Z_2 e^2 (-1)\, r^{-2} A + Be^2 (-n)\, (r)^{-(n-1)}$$

or
$$\left[\frac{d\,(\text{P. E.})}{dr}\right]_{r = r_0} = -\ Z_1 Z_2 e^2 (-1)\, r_0^{-2} A + Be^2 (-n)\, (r_0)^{-(n-1)} = 0$$

or
$$B = \frac{A\, Z_1\, Z_2}{n}\, r_0^{n-1}. \qquad \text{...(4)}$$

Substituting equation (4) into (3), we get

$$(\text{P.E.})_0 = \frac{Ae^2 Z_1 Z_2}{r_0}\left(\frac{1}{n} - 1\right). \qquad \text{...(5)}$$

The lattice energy, U_0, is the amount of energy released when one mole of sodium chloride crystals is formed from gaseous ions which are at infinite separation. Thus, by definition

$$U_0 = -(\text{P.E.})_0 \times N$$

$$= -\frac{NAe^2 Z_1 Z_2}{r_0}\left(\frac{1}{n} - 1\right)$$

$$= -\frac{NAe^2 Z_1 Z_2}{r_0}\left(1 - \frac{n}{1}\right) \quad ...(6)$$

where N is the Avogadro's number. Equation (6) is known as Born's equation.

Conclusion. From equation (6), it can be seen that:

(*i*) The lattice energy U_0 depends upon the charges of the ions, multiple charged ions giving more lattice energy than singly charged ions. This point is illustrated below :

Inoic Solid	*Nature of Ionic Solid*	*U_0 K cal/mole.*
LiF	uni—univalent ionic solid	240
CaF_2	uni—bivalent ionic solid	632
MgS	bi—bivalent ionic solid	777

(Increasing ↓)

(*ii*) The lattice energy U_0 *depends upon the value of* r_0 —the smaller the value of r_0 the more negative being the value of lattice energy. This point is illustrated below:

Ionic Solid	*r_0 (Å)*	*U_0 kJ/mole*
LiF	2.01	—1008
LiCl	2.57	—853
LiBr	2.75	—828
LiI	3.00	—756

(*iii*) The lattice energy depends on the value of the Madelung constant, which depends on the geometrical arrangement of the ions in the lattice.

(*iv*) The lattice energy depends upon the value of the Born exponent *n*, which may be determined from measurements of the compressibilities of ionic crystals.

Correction of Born's Equation. In very accurate calculations some correction factors are made in equation (6) which does not take into account of some minor forces. There are three main correction factors:

(*a*) *Inclusion of van der Wall's Forces.* These forces operate between all atoms ions or molecules, but are relatively very weak. They are due to attractions between oscillating dipoles in adjacent atoms and vary approximately as $1/r^6$. They can be calculated from the polarizabilities and ionisation potentials of the atoms or ions.

(*b*) *Use of a More Rigorous Expression for the Repulsive Energy.* The equation (2) for the repulsive energy is not strictly correct from quantum mechanical considerations. More refined expressions do not greatly change the results, however.

(*c*) *Consideration of Zero Point Energy of the Crystals.* The zero point energy of the crystal is the energy of vibration of the ions which the crystal possesses even at the absolute zero. This can be calculated from the lattice vibration frequencies.

The calculation of lattice energy of an ionic crystal by using equation (6) is very important since, in general, there is no direct way to measure them experimentally although it can be obtained from certain experimental data using the Born-Haber cycle which is discussed immediately below :

8.7. Born-Haber Cycle (Experimental Determination of Lattice Energy)

This cycle devised by **Born** and **Haber** in 1919 relates the lattice energy of a crystal to other thermochemical data. This approach is based upon the assumption that the formation of an ionic crystal may occur either by direct combination of the elements or by an alternative process in which

(i) the reactants are vapourised and converted into gaseous atoms,

(ii) the gaseous atoms are converted into ions, and

(iii) the gaseous ion are combined to give the product.

For a single substance such as an alkali metal halides (MX), such a process might be formulated:

$$\begin{array}{ccc} MX(s) & \xleftarrow{-U_0} & M^+(g) + X^-(g) \\ \uparrow -\Delta H_f & & \uparrow +I \qquad \uparrow -E \\ M(s) + \frac{1}{2}X_2(g) & \xrightarrow{+S \ +/2D} & M(g) + X(g) \end{array}$$

Each step being denoted by an energy quantity. They have the following significances:

(i) **S** : It is the sublimation energy of the metal

$$M(s) \rightarrow M(g)$$

As the energy is required for the process (endothermic), its value is reported as a positive quantity.

(ii) **D** : It is the dissociation energy of X_2 (gas) into atoms.

$$\frac{1}{2}X_2(g) \xrightarrow{l/2D} X(g)$$

This process is also endothermic and its value is considered to be a positive quantity.

(iii) **I** : It represents ionisation energy, *i.e.*, energy required for the removal of an electron from an isolated gaseous atom, M. This process is endothermic and its value is reported as positive quantity.

$$M(g) \xrightarrow{+I} M^+(g) + e$$

(iv) **E** : It represents electron affinity, *i.e.*, amount of energy released when a neutral gaseous atom gains an electron. The process is exothermic and its value is given as negative quantity.

$$X(g) + 1e \xrightarrow{-E} X^+(g)$$

(v) **U_0** : It represents lattice energy, *i.e.*, the amount of energy released when one mole of gaseous positive and negative ions condense to form an ionic solid.

$$M^+(g) + X^-(g) \xrightarrow{-U_0} MX(s)$$

The process is exothermic and its value is always negative.

(vi) ΔH_f : It is the heat of formation which is equal to the amount of energy released when one mole of a substance is formed from its elements.

$$M(s) + \frac{1}{2}X_2(g) \xrightarrow{-\Delta H_f} MX(s)$$

This is single step reaction. Energy is released in the process and its value is reported as a negative quantity.

As the same product is obtained from the same reactants by either path, the total energy changes must be equal in the two paths. It follows that

$$-\Delta H_f = S + \frac{1}{2}D + I - E - U_0 \qquad ...(7)$$

Solving for U_0, that

$$U_0 = \Delta H_f + S + \frac{1}{2}D + 1 - E \qquad ...(8)$$

Thus, lattice energy can be calculated if other thermochemical quantities are known.

Example. *Calculate the lattice energy of KBr. The heat of sublimation of potassium is 88.2 kJ/mole, the heat of dissociation of bromine gas is 193.2 kJ/mole, the ionisation potential of potassium is 415.8 kJ/mole, the electron affinity of bromine is –336 kJ/mole and the formation of solid KBr from its elements is —407.4 kJ/mole.*

Solution. Using the Born-Haber's cycle.

$$KBr(s) \xleftarrow{-U_0} K^+(g) + Br^-(s)$$

$-\Delta H_f$ (K(s) + ½Br₂(g) → KBr(s)); $+I$ (K(g) → K^+(g)); $-E$ (Br(g) → Br^-)

$$K(s) + \frac{1}{2}Br_2(g) \xrightarrow{+S\ +\ 1/2D} K(g) + Br(g)$$

As the same product is obtained from the reactants by either path the total energy changes must be equal in the two paths.

$$U_0 = \Delta H_f + S = \frac{1}{2}D + I - E$$

Here S = heat of sublimation of potassium

= 88.2 kJ/mole

D = heat of dissociation of bromine gas

= 193.2 kJ/mole

or $\frac{1}{2}D$ = 96.6 kJ/mole

I = ionisation potential of potassium
= 415.8 kJ/mole

E = the electron affinity of bromine
= – 336kJ/mole (neglect the negative sign)

ΔH_f = heat of formation of solid KBr from its elements.
= – 407.4kJ/mole (neglect the negative sign).

Substituting these values in equation (80, we get

$$U_0 = 407.4 + 88.2 + 96.6 + 415.8 - 336.0$$
$$= 1008.0 - 336.0 = 672.0 \text{ kJ/mole}$$ **Ans.**

Note. While substituting the various values in equation (8), signs attached with E and Q in the above solved example are not considered because these have already been taken into consideration while deriving the equation (8).

Deviation. For a large number of ionic solids the values of lattice energy calculated from equation (6) agree fairly with those determined experimentally by using Born-Haber cycle equation (8). There are however, many ionic solids for which U_0 calculated by equation (6) are much different from each other. Fajan (1924) explained this difference by suggesting that ideal ionic bond is found in a very few compounds. Most of the ionic compounds have some covalent character. The difference between lattice energy calculated by using equations (6) and (8) may be attributed to some amount of covalent character in the ionic solids which arises due to the polarisation of the ions.

Importance of Born-Haber Cycle. (i) The most important application is in the determination of electron affinities which are otherwise difficult to determine by other means.

(*ii*) It is also useful in analysing and correlating the stability of various ionic solids.

(*iii*) It also provides an explanation for why most metals fail to form stable ionic compounds in low valence states such as AlO, MgCl.

8.8. Solid State Defects

An ideal crystal is one that possesses the same unit cell having the same lattice points across the whole of the crystal. When temperature is approaching absolute zero, crystals would possess the tendency to have perfectly ordered arrangement of ions and hence there will be no defects. When the temperature is increased, the chance that a lattice site may not be occupied by an ion will increase. This constitutes a defect. Thus the term defect may be defined as the departure from regularity in the arrangement of the particles (atoms, ions and molecules) constituting a crystal.

The number of defects depends on the temperature. Therefore, they are sometimes known as thermodynamic defects. Mathematically the number of defects formed by cm^3, n, has been given by

$$n = Ne^{-w/2RT}$$

where N denotes the number of sites per cm^3 of the crystal, w denotes the work (energy) required to produce a defect and T denotes the temperature in Kelvin.

Defects in crystals alter their physical properties especially electrical conductivity and diffusion to a great extent and sometimes even their chemical properties but to lesser extent.

Types of Defects. These are of two types :

(*a*) Point defects.

(*b*) Non-stoichiometric defects.

We shall discuss these defects one by one.

(*a*) A Point Defect. This type of defect may arise when any of the constituent particles has been either missing from the crystal lattice or has been dislocated to a position which is meant for another particle or has been got shifted to an interstitial position.

In point defects, there is no change in the electrical neutrality of the crystal and also its stoichiometry is not altered. These defects are of two types :

(*i*) Schottky defect. This defect may arise when some of the lattice points (a cation and an anion) are unoccupied. These unoccupied points are called lattice vacancies or holes (Fig. 8.11 (a). This sort of defect occurs in compounds with high coordination numbers, and where the positive and negative ions have similar size. For example, NaCl and CsCl are ionic solids that have Schottky defect.

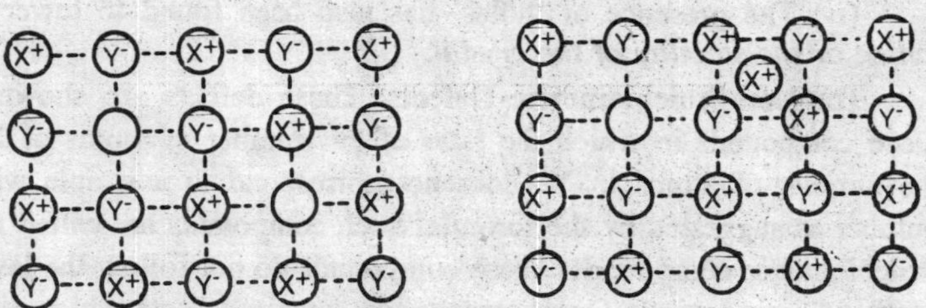

Fig. 8.11 (*a*) Schottky defect of crystals; (*b*) Frenkel defect of crystals.

Schottky defect is called after the German scientist, Schottky who had discovered it in 1930.

(*ii*) **Frenkel defect.** This defect may arise if an ion is not completely missing but has been shifted to an interstitial position in the lattice, thus, leaving a vacancy in its own position Fig. 8.11 (b).

A hole is thus created in the lattice as shown. This defect was discovered by the Russian scientist Frenkel in 1926.

Examples of Frenkel defect are ZnS and AgBr. The Frenkel defect has been generally shown by the following types of compounds :

(*i*) that possess low coordination numbers,

(*ii*) that possess ions of different sizes, and

(*iii*) that possess a highly polarising cation and an easily polarisable anion.

Though vacancies have been created in both Schottky and Frenkel defects, the former results in a decrease in the overall density of the crystalline substance whereas the latter does not.

Consequences of stoichiometric defects. Schottky and Frenkel defects in crystal lead to the following results:

(*i*) The closeness of similar charges brought about by the Frenkel defect leads to an *increase in the dielectric constant of the crystals.* The density of the medium however remains unchanged.

(*ii*) A consequence of both types of defects is that the *crystal is able to conduct electricity to a small extent,* by an ionic mechanism. When an electric field is applied to a crystal having stoichiometric defects, a nearby ion moves from its lattice site to occupy a hole. This results the creation of a new hole and thus another ion moves into it, and so on: thus, the ion migrates from one end to the other, causing the electricity to flow across the whole of the crystal.

(*iii*) The presence of 'holes' has been found to *lower the density of the crystal.*

(*iv*) The presence of 'holes' has also been found to lower the *lattice or the stability of the crystal.*

(b) Non-Stoichiometric Defects. These defects are shown by those compounds in which the ratio of the number of atoms of X-to the number of atoms in Y^- does not correspond to a simple whole number as suggested by the formula. Such compounds are called non-stoichiometric compounds. These compounds do not follow the law of constant composition. These compounds contain defects which are in addition to the usual thermodynamic (Schottky and Frenkel) defects.

Non-stoichiometric defects, have been found to be of two types, depending upon whether there is excess of positive share or excess of negative charge. These are termed as metal excess defect and metal deficiency defects, respectively.

(a) **Metal excess defects.** When positive charge is in excess, this is known as metal excess defect. This defect may arise in two ways :

(*i*) There may be missing of negative ion from its lattice site leaving a hole which will be occupied by an extra electron so as to maintain the electrical balance [Fig. 8.12 (a)]. Thus, there would be an excess of positive (metal) ions, although the crystal, as a whole, is neutral.

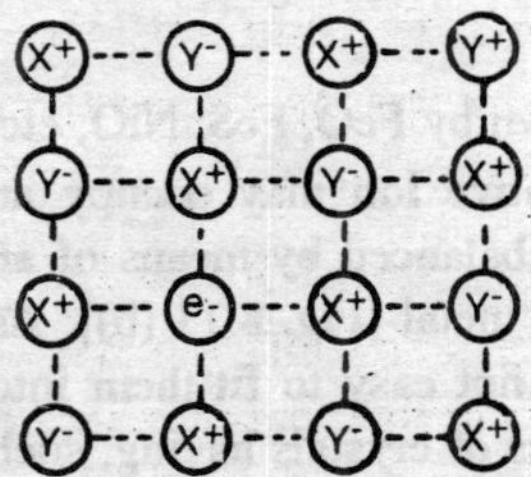

Fig. 8.12 (a) Metal excess defect caused by a missing ion.

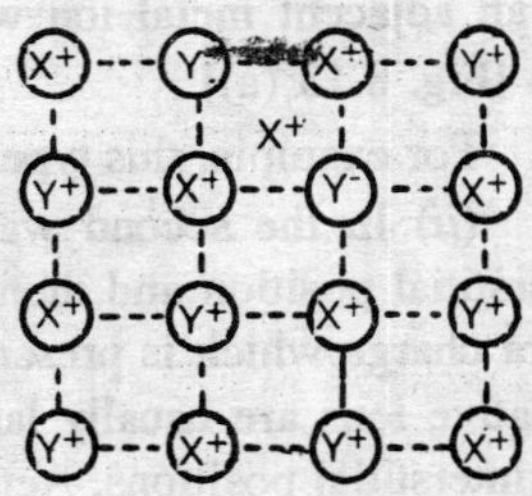

Fig. 8.12 (b) Metal excess defect due to extra positive ion.

This type of defect is generally not very common. For example, NaCl when treated with sodium vapour gives a yellow non-stoichiometric form of NaCl which has excess of sodium ions. In such a case, the extra positive charge due to extra sodium ion, has been balanced by the presence of free electron in the lattice.

(*ii*) The metal excess defect may also appear in another way by occupying an interstitial position in the lattice with an extra positive ion and to maintain electrical balance, an electron will be present in the interstitial space [Fig. 8.12 (b)]. For example, this defect is shown in zinc oxide crystal.

Consequences of metal excess defects

(*i*) As the crystals associated with metal excess defects of the first or second kind contain free electrons, they can conduct electricity to some extent because the number of defects and hence the number of electrons are very small. Such crystals are generally termed as *semi-conductors.*

(*ii*) The crystals showing metal excess defects are generally coloured. This due to the presence of free electrons. When these electrons are excited to higher energy levels by absorption of certain wavelengths from the visible white light, these compounds appear coloured. For example, zinc oxide is white is cold but appears yellow when hot.

(b) **Metal deficiency defects.** Theoretically, metal deficiency defects may arise in two ways. However, both require variable valency of the metal and these are expected to occur in the transition metals. These may arise in one of the two ways :

(*i*) In the first way, there occurs the missing of a positive ion from its lattice site and then the exta negative charge will be balanced by an adjacent metal ion which has two positive charges instead of one [Fig. 8.13 (a)].

For example, this type of defect is shown by FeO, FeS, NiO, etc.

(*ii*) In the second way, an extra negative ion may occupy an interstitial position and then the charges are balanced by means of an extra charge which is present on an adjacent metal [Fig. 8.13 (b)]. As negative ions are usually large, it would be not easy to fit them into the interstitial positions. Actually no examples of crystals having, such negative interstitial ions have been reported so far.

Crystal having metal deficiency defects are generally semiconductors.

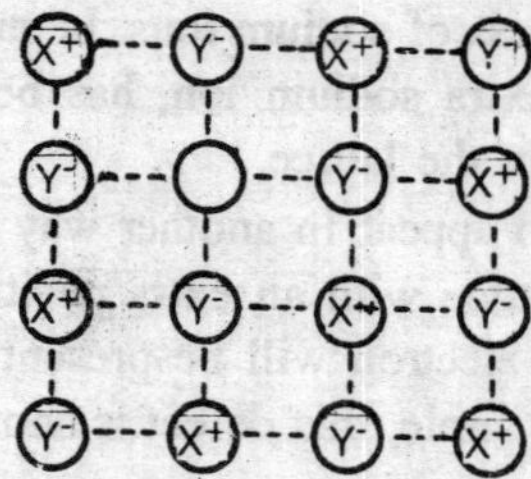

Fig. 8.13. (a) Metal deficiency defect due to a missing positive ion.

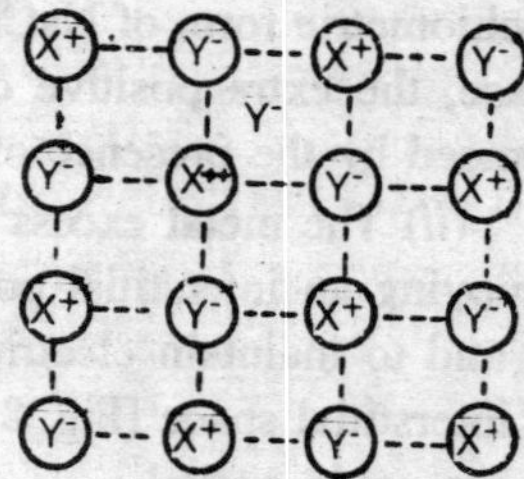

Fig. 8.13. (b) Metal deficiency defect due to an extra negative ion.

Consequences of metal deficiency defects

Crystals with metal deficiency defects are semi-conductors. This property arises from the movement of an electron from one ion to another ion. The substances permitting this type of movement are known as *p*-type semi-conductors.

8.9. Sintering

Sintering is an important technological process in which powder particles can be bound together at a temperature lower than the melting point to impart desirable properties to the product. Many industrial ceramics—magnetic materials like ferrites and ferroelectrics like barium titanate/lead zirconate—are synthesized using sintering as one of the stages.

The driving force for sintering is the large surface free energy associated with the fine-gained powders. Therefore, there is always a tendency to minimize the free energy by spontaneously decreasing the particle surface. However, in practice, high temperatures are usually required at which sintering rates become appreciable. Besides temperature, sintering rate also depend upon various factors like particle size, distribution of sizes, non stoichiometry, impurities, enveloping gas atmospheres, magnetic field (for magnetic ceramics), electric field (for ferroelectric ceramics), etc. The sintering process changes properties like mechanical hardness, thermal conductivity, ductility, transparency, coervice force, magnetic permeability, energy product, electric polarization, etc., and therefore, the experimental conditions have to be controlled to obtain desired properties.

Obviously, sintering is effected through mass transport involving atomic/ionic movements which are usually of the order of the particle size. Frenkel was the first to suggest a model for describing the various stages of sintering; since then a large number of workers have contributed to this area due to its technological importance. The mass transport can be due to (a) diffusion, (b) flow or (c) evaporation-condensation.

8.10. Liquid-Crystals

Some substances melt to form a transition phase which has both a considerable degree of the internal order associated with crystals and can also flow like liquids. **This transition state lies between the complete long range order of a crystal and the almost long ranged disorder of a liquid.** A liquid crystal shows the mechanical properties of liquids and optical properties of crystals. The liquid crystals interact with plane polarized light and also diffract light in accordance with Bragg's law : $n\lambda = 2d \sin\theta$.

Some solids like **ammonium-cis-octadec-9-enoate and cholesterol esters** form liquid crystals. Their molecules are long and assymetric.The long molecules forming liquid crystals are either flat (*plate like*) or cylindrical (*rod like*) in shape. Fig. 8.14 (*a*) shows the ordered arrangement of long molecules in three dimension in a crystal. Fig 8.14 (*b*) shows arrangement of molecules in a smetic liquid crystal. The molecules group together in well defined planes which are equi-distant from each other and the molecules lie perpendicular to these planes. The molecules do not move from one plane (*1 layer*) to another but can reorient within a plane (*layer*) and have some disorder. These planes (*layers*) can slide past each other giving the liquid crystal a fluid character. In a **Nematic liquid crystal,** the molecules in a plane (*layer*) are not equispaced [Fig. 8.14 (*c*)] although they are still parallel to each other (*same orientation*) and have greater disorder.

. The parallel molecules can pass from one plane into another. In **a cholesteric liquid crystal,** the molecules lie lengthwise and parallel in planes. These liquid crystals are said to be in **para crystalline state.** They diffract light because the planes (*layers*) are static. Fig. 8.14 (*d*) shows the disorder in a liquid.

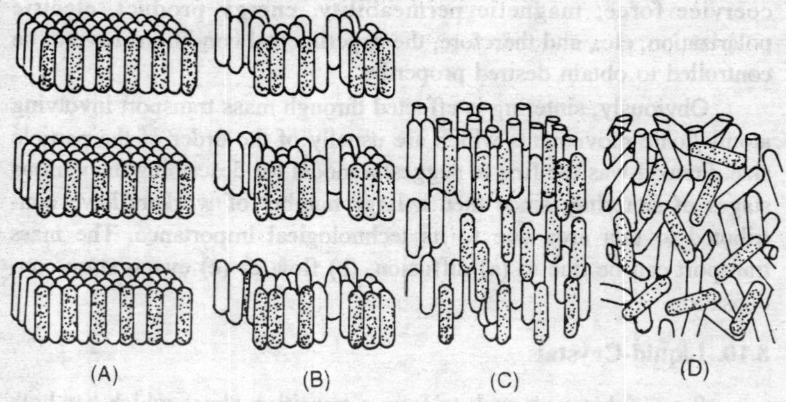

Fig. 8.14 Paracrystalline states of a liquid crystal.

"The different paracrystalline states of a liquid crystal are highly sensitive to slight temperature changes and electric current. These paracrystalline states usually have different colours. A slight change in temperature or electric current changes the colour of a liquid crystal because of the orientation of molecubes. Not only this the colour can be restored by reversing the change. The liquid crystals have many practical applications. They are now widely used to display numbers in electronic calculators, digital watches and thermometers because a little energy (weak electric field) is used up. They can also be used to locate structural flaws in a metal. They are also used to provide sensitive tests for air pollutants and to locate veins which have slightly lower temperature than skin. A recent work shows that portions of cell membranes and organelle membranes are liquid crystals. Japan has produced colour Televisions using liquid crystals instead of the high voltage cathode ray tube."

8.11. Solid State Reactions

Reactivity of solids is a subject of both technological and fundamental importance. Sometimes we may be interested in high reactivity, e.g., in heterogeneous catalysis and solid propellants, and sometimes in low reactivity, e.g., in corrosion of metals and alloys. In solid state electronics, we have to carry out reactions preferentially at certain chosen parts. Therefore, it is imperative to understand the basic mechanisms of reactions involving solids, so as to be able to control reaction rates in specified situations.

Systematic studies on reactions between solids were initiated by J.A. Hedvall and G. Tammann around 1922-25. Essential conclusions of their investigations were as follows:

(i) Solid-solid reactions are exothermic.

(ii) Direct reaction between grains of reactants takes place during heating. Melting and/or vaporization processes do not play any role in the reaction.

(iii) If one of the reactants undergoes a polymorphic transition at say, T_c, then the reaction proceeds at an appreciable rate at T_c.

Although some of their findings were found to be at variance with the results of subsequent investigations, their work undoubtedly provided a great impetus to further researches on reactions between solids. Their observation that liquid or gaseous phases do not play any role in the solid state reactions was criticised by many workers. It was postulated that reaction between solid grains can proceed only after the abscence of participation in reaction by liquid and/or gaseous phases/species is established. Local heating effects may melt a part of at least one of the reactants, which will then react with the other reactant(s). The heat evolved will melt some more of the earlier reactant the reaction rate may soon become appreciable. Similarly, if it is found that the reaction rate is high at temperatures much lower than the dissociation or sublimation temperature of a reactant, when measured in isolation, it does not prove conclusively that dissociation or sublimation had not occurred in the presence of other reactants. Around 1930, W. Jander, C. Wagner and W. Schottky made notable contributions to thermodynamics and kinetics of solid-state reactions. Wagner proposed that diffusion of ions and/or electrons through the lattice(s) of reactants is responsible for solid-solid reactions.

Wagner's Theory

For ionic compounds, it is assumed that the mobility of cations is relatively high, while the anions are immobile and do not participate in the diffusion process. If more than one cations are diffusing, the difference in mobilities sets up a diffusion potential which then controls the rate of migration; the mobility of the fast moving ions is slowed down, while that of the slow moving ions is increased, so that one may speak of an average migration rate of the ions. Wagner illustrated his ideas through his classical experiment on the reaction between silver and sulphur to form $\alpha-Ag_2S$.

A silver block was separated from sulphur by two blocks of silver sulphide ($\alpha-Ag_2S$) as shown in Fig. 8.15. The temperature was raised to about 220° C and maintained there for about an hour. Sulphur melted and reacted with silver metal through the Ag_2S blocks; that the reaction took place was shown by the loss in weight of the silver block and gain in weight of $\alpha-Ag_2S$(II) block. The changes in weight in conjuction with the advance of $\alpha-Ag_2S$(II)/S interface into molten sulphur, indicate that (i) Ag^+ ions must have diffused through the

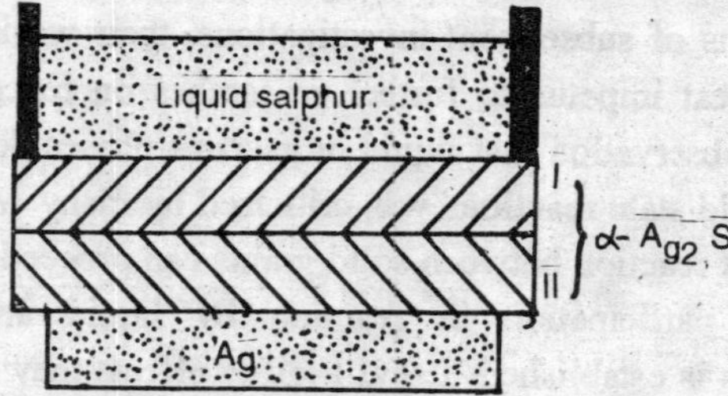

Fig. 8.15. Wagner's experiment showing mechanism of reaction for the sulphiding of silver.

$\alpha-Ag_2S$ blocks and (ii) reaction between Ag^+ and S^{2-} ions occurred at the $\alpha-Ag_2S$(II)/S interface. Wagner suggested the following mechanism for the reaction :

(i) Since the conductivity of $\alpha-Ag_2S$ is mixed, i.e., ionic and electronic, Ag+ ions diffused through the alpha-Ag_2S blocks and arrived at the $\alpha-Ag_2S$(II)/S interface; to maintain charge neutrality, electrons also diffused in the same direction as the Ag^+ ions. The sulphide (S^{2-}) ions were immobile and, therefore, did not diffuse.

(ii) The electrons converted the sulphur atoms adsorbed on the $\alpha-Ag_2S$(II) surface into sulphide (S^{2-}) ions, which built up a framework of the anion lattice.

(iii) The Ag+ ions arriving at the $\alpha-Ag_2S$(II)/S interface occupied the cation sites in the newly-formed anion lattice.

(iv) The formation of $\alpha-Ag_2S$ increased the weight of $\alpha-Ag_2S$(II) block and the $\alpha-Ag_2S$(II)/S interface advanced into the molten sulphur.

Wagner's mechanism appeared to be applicable in the case of addition reactions leading to the formation of spinels and silicates. When magnesium oxide (mgO) reacts with alumina (Al_2O_3), diffusion of Mg^{2+} and Al^{3+} ions takes place in opposite directions, while the larger oxide (O^{2-}) ions are immobile. The reactions occurring at the interfaces are :

$$MgO/MgAl_2O_4 : 4MgO - 3Mg^{2+} + 2Al^{3+} \rightarrow MgAl_2O_4$$

$$MgAl_2O_4/Al_2O_3 : 4Al_2O_3 - 2Al^{3+} + 3Mg^{2+} \rightarrow 3MgAl_2O_4$$

For the reaction,

$$MgO + MgSiO_3 \rightarrow Mg_2SiO_4$$

magnesium (Mg^{2+}) and silicon (Si^{4+}) ions diffuse in opposite directions, leading to the following reactions at the interfaces :

$$MgO/Mg_2SiO_4 : 4MgO - 2Mg^{2+} + Si^{4+} \rightarrow Mg_2SiO_4$$

$$Mg_2SiO_4/MgSiO_3 : 4MgSiO_3 - Si^{4+} + 2Mg^{2+} \rightarrow 3Mg_2SiO_4$$

Wagner's mechanism may also be applied to the following reactions :

$$2AgI + HgI_2 \rightarrow Ag_2HgI_4$$

$$AgCl + NaI \rightarrow AgI + NaCl$$

In situations where movement of cations is improbable, the diffusion of anions may be invoked. This will be especially possible when a high concentration of anion vacancies exists. Anion vacancies may be produced extrinsically by, say, incorporation of aliovalent anions into the lattice. In general, it may be stated that a relationship exists between the diffusion rate of ions and ionic conductivity because both are influenced by the presence of ionic vacancies. Also, 'open' structures favour fast diffusion of ions, e.g. diffusion of Ag^+ ions in silver halide latices. The state of aggregation, e.g., power or single crystal, will also influence reactivity. Since vacancies and interstitials can be artificially produced through irradiation, it can be considered as another factor capable of influencing reactivity.

In conclusion, it may be mentioned that Wagner's theory has limited applications and cannot be applied to all solid-solid reactions.

9

Solved Problems

Q. 1. For the cubic unit cell shown in the figure, how many O atoms and how many ● atoms are there per unit cell?

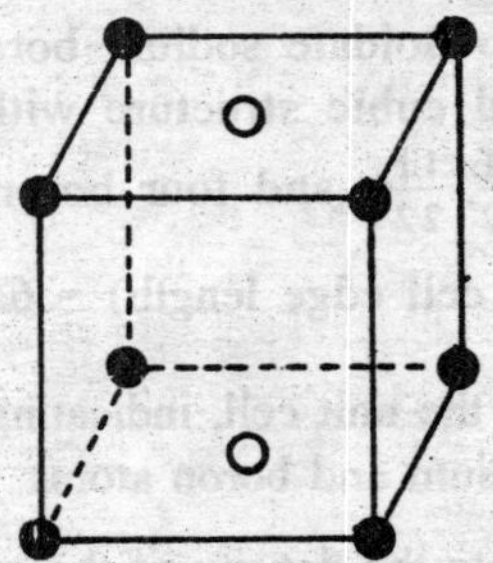

Fig. 9.1

Q. 2. The Weiss indices of a certain crystal face are $\frac{1}{2}, \frac{2}{3}, \infty$. What are the corresponding Miller indices?

Q. 3. For the crystal face shown in the figure, what are the Weiss indices and what are the Miller indices?

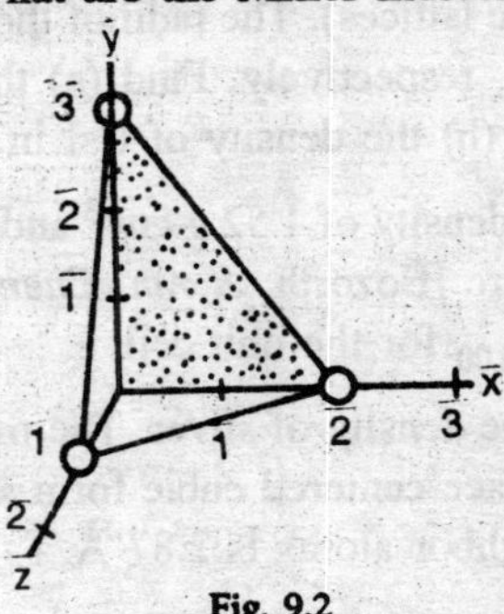

Fig. 9.2

Q. 4. X-ray studies indicate that potassium silyl, $KSiH_3$, formula weight 70.18, has a cubic structure of the NaCl type, with potassium at positions at 000, $\frac{1}{2}\frac{1}{2}0$, $0\frac{1}{2}\frac{1}{2}$, $\frac{1}{2}0\frac{1}{2}$ and silicon positions at $00\frac{1}{2}$, $\frac{1}{2}00$, $0\frac{1}{2}0$, $\frac{1}{2}\frac{1}{2}\frac{1}{2}$. The edge of the unit cell is 7.15 Å in length. No effort was made to locate the hydrogen atoms, all of which are held in the form of the silyl anion, SiH_3.

(a) Sketch a representative unit cell showing the relative positions of the potassium and silicon atoms.

(b) Calculate the distance in Ångstrom units between nearest potassium-silicon neighbors.

(c) Calculate the density of the crystal.

Q. 5. According to Soldate sodium borohydride, $NaBH_4$, has a face-centered cubic structure with four sodium atoms at 000, $0\frac{1}{2}\frac{1}{2}$, $\frac{1}{2}0\frac{1}{2}$, $\frac{1}{2}\frac{1}{2}0$, and four boron atoms at $\frac{1}{2}\frac{1}{2}\frac{1}{2}$, $\frac{1}{2}00$, $0\frac{1}{2}0$, $00\frac{1}{2}$. a_o (unit cell edge length) = 6.15 Å.

(a) Sketch the unit cell, indicating the relative positions of the sodium and boron atoms.

(b) Calculate the density of the crystal.

Q. 6. Cesium bromide crystallizes in the cubic system. Its unit cell has a Cs^+ ion at the body center and a Br^- ion at each corner. Its density is 4.44 g cm^{-3}. Determine (a) the length of the unit cell edge, and (b) the d_{200} distance.

Q. 7. CsI has the same structure as CsCl (two interpenetrating simple cubic lattices). The radii of the Cs^+ and I^- ions are 1.69 and 2.16 Å, respectively. Find (a) the volume of a unit cell in CsI, and (b) the density of CsI in g cm^{-3}.

Q. 8. KCN has a density of 1.52 g cm^{-3} and crystallizes in the NaCl-type structure [Bozorth, *J. Am. Chem. Soc.*, **44**, 317 (1922)]. Calculate d_{100} for the unit cell.

Q. 9. Calculate the density of silver. The metal is known to crystallize in the face-centered cubic form and the distance between nearest neighbor atoms is 2.87 Å.

Q. 10. The silver perchlorate-benzene complex, $AgClO_4 \cdot C_6H_6$, is orthorhombic with unit cell dimensions a_0 = 7.96, b_0 = 8.34, and c_0 = 11.7 Å. The formula weight is 285 and there are four molecules per unit cell [Rundle and Goring, *J. Am. Chem. Soc.*, 72, 533⁻ (1950)]. Calculate the density of the crystal.

Q. 11. Lithium borohydride, $LiBH_4$, has an orthorhombic structure with four molecules per unit cell and unit cell dimensions of a_0 = 6.81, b_0 = 4.43, and c_0 = 7.17 Å. Calculate the density of this crystal.

Q. 12. A certain organic compound crystallizes in the orthorhombic system. The unit cell has edges of 12.05, 15.05, and 2.69 Å. There are two molecules per unit cell. The density of the crystal is 1.419 g cm^{-3}. Find the molecular weight of the compound.

Q. 13. What is the physical significance of the *n* (the "order" of the X-rays) in the Bragg equation, $n\lambda = 2d \sin \theta$?

Q. 14. The crystal structure of potassium cyanide was studied by Bozorth [*J. Am. Chem. Soc.*, 44, 317 (1922)], who found the following angles of reflection from the 100 planes :

Angle of reflection	5°23′	10°51′	4°47′	4°40′
Order of reflection	2	4	2	2
Wavelength, Å	0.614	0.614	0.545	0.534

The structure is cubic. Index the lines and indicate the type of cubic lattice. The following table will be helpful:

hkl	100	110	111	200	210	211	220	221, 300
$h^2 + k^2 + l^2$	1	2	3	4	5	6	8	9
hkl	310	311	222	320	321	400		
$h^2 + k^2 + l^2$	10	11	12	13	14	16		

Q. 15. LiBr, NaBr, KBr, and RbBr all have the same crystal structure. X-ray diffraction, however, indicates that RbBr has a simple cubic lattice, and that the other three have face-centred lattices. Explain.

Q. 16. Explain why a crystal cannot have an axis of greater than sixfold symmetry.

Q. 17. List the symmetry elements of the solid shown in the figure. *bcd* is an equilateral triangle; the other three faces are congruent isosceles triangles.

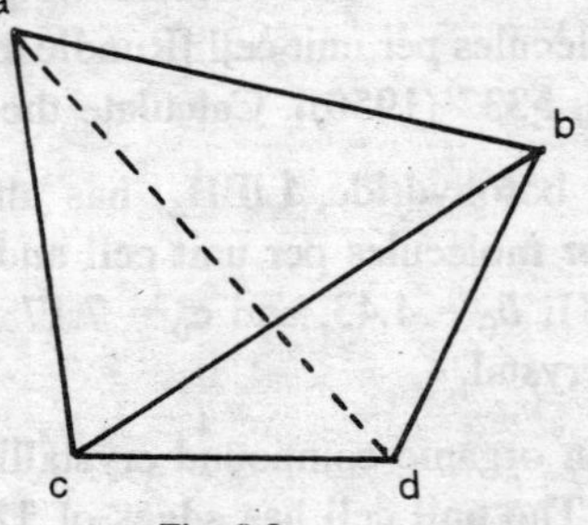

Fig. 9.3

Q. 18. List the symmetry elements of the crystal shown in the figure.

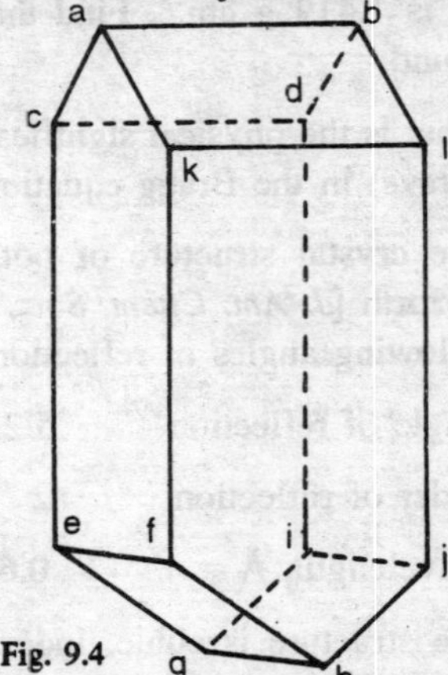

Fig. 9.4

Q. 19. How many twofold axes of symmetry are possessed by the crystal shown in the figure? Sketch them in. The crystal is a perfect cube except for the beveled corners.

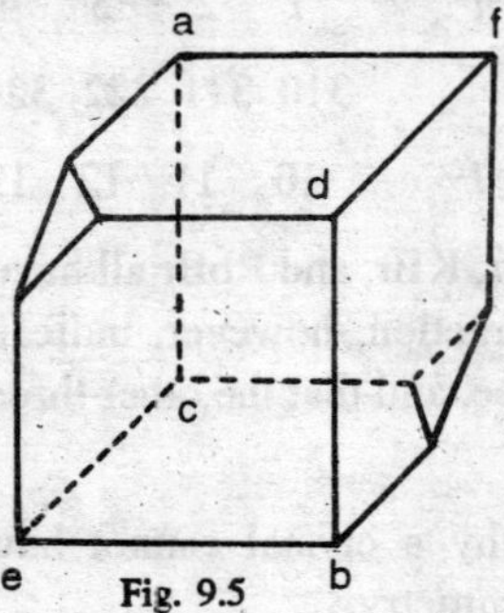

Fig. 9.5

Q. 20. Show that a twofold axis of rotary inversion ($\bar{2}$) is equivalent to a mirror plane.

Q. 21. *(a)* Sketch a two-dimenstional lattice of closest-packed identical circles, indicating a suitable unit cell.

(b) Find the fractional void area in this lattice. (For a triangle with sides a, b, c, the area is $[s(s—a)\ (s—b)\ (s—c)\ ]^{1/2}$, where

$$s = 1/2\,(a + b + c).$$

Q. 22. Three uni-univalent ionic crystals, AX, AY, and AZ are composed of ions having the following radii (in arbitrary units):

A^+	X^-	Y^-	Z
1.00	1.00	2.00	3.00

Assume that the ions are hard spheres.

(a) Predict whether each crystal will have the sodium chloride or the cesium chloride structure. Explain.

(b) Assuming that your predictions are correct, find the volume of the unit cell in each of the three crystals.

Q. 23. Using the following ionic radii, find the fractional void volume in *(a)* CsCl, *(b)* NaCl, *(c)* LiCl. Assume that the ions are hard spheres in contact.

Ion	*Radius, A*
Li+	0.60
Na+	0.95
Cs+	1.69
Cl	1.81

In CsCl the positive ions and the negative ions each comprise a simple cubic lattice. Each ion has eight oppositely charged ions as nearest neighbors. In NaCl and LiCl the ions of each sign comprise a face-centred cubic lattice. Each ion has six oppositely charged ions as nearest neighbors.

SOLUTION

A-1. The unit cell contains 8 • atoms, each shared with 8 other unit cells, so the net number of • atoms that can be assigned to a unit cell is $8 \times \frac{1}{8} = 1$. The unit cell contains 2 • atoms, each one shared between two unit cells, so the net number of • atoms that can be assigned to a unit cell is $2 \times \frac{1}{2} = 1$.

A-2. To determine the Miller indices we take the reciprocals of the Weiss indices, giving 2/1, 3/2, 0, and then multiply the latter by the denominator 2 to eliminate the fractions, giving 4, 3, 0.

A-3. The Weiss indices are the coordinates of the intercepts of the plane, 2, 1, 3. To obtain the Miller indices, take the reciprocals of the Weiss indices, giving 1/2, 1/1, 1/3 and then multiply through by 6 to eliminate the fractions, giving 3, 6, 2.

A-4. (*a*) See figure. 0 = K, • = Si. For clarity the 100, 010, and 001 faces are not shown in the drawing. The K atoms are at the corners and face centres. The Si atoms are at the body center and at the edge centers.

(*b*) Nearest K—Si neighbors are 7.15/2 = 3.57 Å apart.

(*c*) Volume of unit cell = $(7.15 \times 10^{-8})^3 = 365 \times 10^{-24}$ cm^3.

Number of unit cells per gram formula weight $KSiH_3$ = $\frac{1}{4} \times 6.023 \times 10^{23}$. Volume occupied by 1 g formula weight $KSiH_3 = \frac{1}{4} \times 6.023 \times 10^{23} \times 365 \times 10^{-24} = 54.9$ cm^3.

Density = 70.18/54.9 = 1.28 g cm^{-3}.

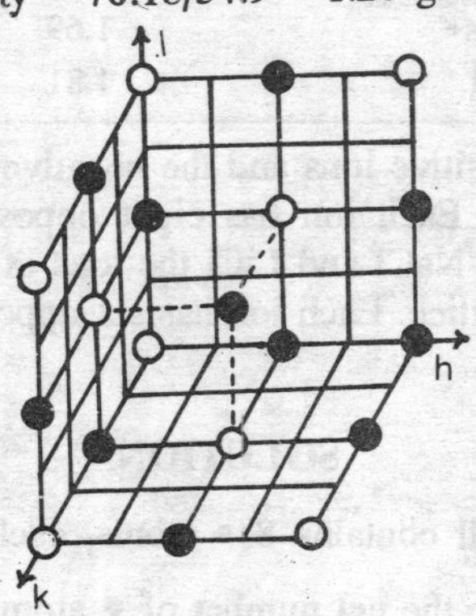

Fig. 9.6

A-5. (*a*) See figure.

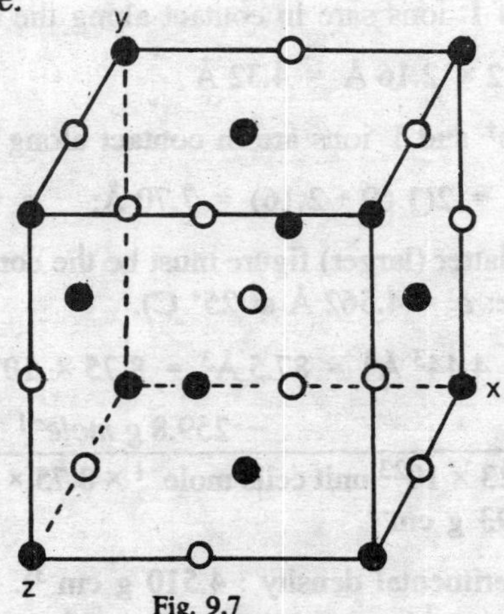

Fig. 9.7

(*b*) Formula weight of $NaBH_4$ = 37.8

Na atoms at corners = 8 × 1/8 = 1

Na atoms at face centers = $6 \times \frac{1}{2} = 3$

Total: 4 Na atoms per unit cell

B atoms at edge centers = $12 \times \frac{1}{4} = 3$

B atom at body center = 1 × 1 = 1

Total : 4B atoms per unit cell

Volume of unit cell $= (6.15 \times 10^{-8})^3\ cm^3 = 233 \times 10^{-24}\ cm^3$

Volume of 1 g formula weight of unit cells =

$\frac{1}{4} \times 6.023 \times 10^{23} \times (6.15 \times 10^{-8})^3\ cm^3 = 35.1\ cm^3$

Density = 37.8 g/35.1 cm^3 = 1.08 g cm^{-3}

A-6. (*a*) The unit cell contains $1Cs^+$ and $1Br^-$.

$$\frac{212.81\ g\ mole^{-1}}{4.44\ g\ cm^{-3} \times 6.023 \times 10^{23}\ \text{unit cells mole}^{-1}}$$

$= 7.97 \times 10^{-23}\ cm^3\ (\text{unit cell})^{-1} = e^3$

$e = \sqrt[3]{79.7 \times 10^{-24}}\ cm = 4.30 \times 10^{-6}\ cm = 4.30\ Å$

(*b*) $d_{200} = \frac{1}{2} \times d_{100} = \frac{1}{2} \times 4.30\ Å = 2.15\ Å$

A-7. *(a)* If I^- ions sare in contact along the edge,

$e = 2 \times 2.16\ Å = 4.32\ Å$

If Cs^+ and I^- ions are in contact along the body diagonal,

$e\sqrt{3} = 2(1.69 + 2.16) = 7.70\ Å; \quad e = 4.44\ Å$

The latter (larger) figure must be the correct one (experimental value: $e = 4.562\ Å$ at 25° C).

$e^3 = 4.44^3\ Å^3 = 87.5\ Å^3 = 8.75 \times 10^{-23}\ cm^3$

(b) $$\frac{259.8\ g\ mole^{-1}}{6.023 \times 10^{23} \text{ unit cells mole}^{-1} \times 8.75 \times 10^{-23}\ cm^3 (\text{unit cell})^{-1}}$$
$= 4.93\ g\ cm^{-3}$

(experimental density : 4.510 g cm^{-3}).

A-8. d_{100}^3 = volume per unit cell; formula weight : 65.11

$$d_{100}^3 = \frac{(65.11/1.52)\ cm^3\ mole^{-1}}{6.023 \times 10^{23} \times \frac{1}{4} \text{ unit cells mole}^{-1}}$$

$$= 284 \times 10^{-24}\ cm^3\ mole^{-1}$$

$$d_{100} = \sqrt[3]{284 \times 10^{-24}} = 6.57 \times 10^{-8}\ cm = 6.57\ Å$$

A-9. The unit cell contains 4Ag atoms ($\frac{1}{8} \times 8 + \frac{1}{2} \times 6 = 4$). The nearest neighbors are indicated in the figure. From this diagram it can be seen that 2.87 Å $= \frac{1}{2} \cdot \sqrt{2} \times d_{100}$. Therefore

$d_{100} = 2.87/0.707 = 4.06\ Å$

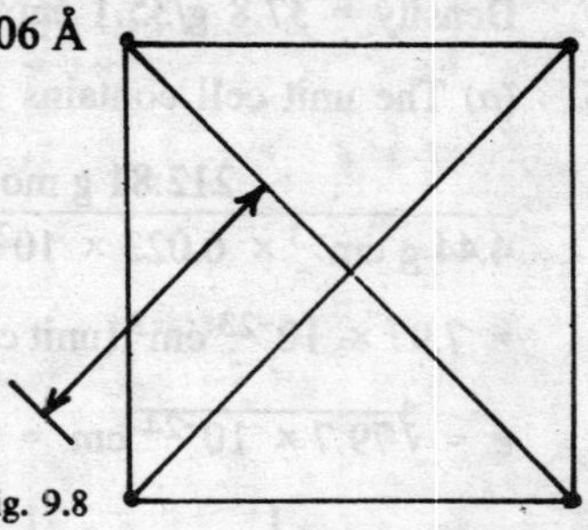

Fig. 9.8

$$V = \frac{1}{4} \times 6.023 \times 10^{23} \times (4.06 \times 10^{-8})^3$$

$$= 10.08\ \text{cm}^3\ (\text{g-atom})^{-1}$$

$$\text{Density} = \frac{107.87}{10.08}\ \frac{\text{g (g-atom)}^{-1}}{\text{cm}^3\ (\text{g-atom})^{-1}} = 10.7\ \text{g cm}^3$$

A-10. Volume occupied by 1 mole of complex is

$$V = 7.96 \times 8.34 \times 11.7 \times (10^{-8})^3 \times \frac{6.023}{4} \times 10^{23}\ \text{cm}^3 = 117\ \text{cm}^3$$

Mass of 1 mole of complex is M = 285 g; density = M/V = 2.44 g cm^{-3}; experimental value, 2.4.

A-11. Volume occupied by 1 mole of $LiBH_4$ = V.

$$= \frac{1}{4} \times 6.81 \times 4.43 \times 7.17 \times (10^{-8})^3 \times 6.023 \times 10^{23}\ \text{cm}^3$$

$$= 32.6\ \text{cm}^3$$

Mass of 1 mole = M = 21.76 g; density = M/V = 0.668 g cm^{-3}

A-12. Volume per molecule = 12.05 × 15.05 × 2.69

$$= 488\ \text{Å}^3\ (\text{unit cell})^{-1}$$

$$= 244\ \text{Å}^3\ \text{molecule}^{-1}$$

$$= 2.44 \times 10^{-22}\ \text{cm}^3\ \text{molecule}^{-1}$$

$$\text{Molecular weight} = 1.419\ \text{g cm}^{-3} \times 2.44 \times 10^{-22}\ \text{cm}^3\ \text{molecule}^{-1}$$

$$\times\ 6.023 \times 10^{23}\ \text{molecule mole}^{-1}$$

$$= 209\ \text{g mole}^{-1}$$

A-13. *n* (the "order" of the X-rays) is the number of wavelengths of the x-rays between the given atomic planes.

A-14.

θ	$\sin^2\theta$	$\sin^2\theta \sin^2 (4.40')$	h^2	k^2	l^2	*hkl*
5 23′	0.00881	1.33 = 4/3		4		200
10 51′	0.03545	5.35 = 16/3		16		400
4° 47′	0.00696	1.05 = 3/3		3		111
4 40′	0.00662	1 = 3/3		3		111

The calculated *hkl* values are all odd or all even. Therefore the structure is face-centred cubic. Actually it is the NaCl structure—two interpenetrating face-centered cubic lattices.

In general,

simple cubic	h, k, l may have any values
face-centred cubic	h, k, l are all odd or all even, including zero
body-centred cubic	$h + k + l$ must be even

A-15. Rb^+ and Br^- have the same number of electrons (36). Thus they are practically equivalent for the purpose of reflecting X-rays. A RbBr crystal appears to be simple cubic if the two kinds of ion are treated as identical. In the other three crystals the two kinds of ions have different numbers of electrons. Each kind of ion occupies a face-centred cubic lattice. The x-rays "see" each of these two lattices (which have identical dimensions) and the result is a diffraction pattern characteristic of a face-centred lattice with variations of intensity resulting from interference between the reflections from the two lattices. (For RbBr the effect of these interferences is to give a pattern identical with that of a simple cubic lattice).

A-16. Any axis of symmetry must have a plane of lattice points perpendicular to it. The overall lattice plane formed by all the adjacent sets of corresponding lattice points must consist of regular polygons that fit together exactly. This is possible only when the symmetry axis is 3-, 4-, or 6-fold.

A-17. One 3-fold axis (*a* to center of face *bcd*). There reflection planes (through *ab* and midpoint of *cd*, etc.).

A-18. One $\bar{4}$ axis (4-fold improper axis, or 4-fold axis of rotary inversion) from midpoint of *ab* to midpoint of *gh*.

One 2 axis (twofold proper axis) from midpoint of *ab* to midpoint of *gh*.

One plane of symmetry containing the *ab* and the midpoint of line *gh*.

One plane of symmetry containing line *gh* and the midpoint of line *ab*.

A-19. There are three. Referring to the figure suitable answers are line *ab*, line *cd*, line *ef*. The diagonal drawn through the beveled corners is not acceptable.

A-20. As can be seen in the figure the 2-fold rotation of ① to ② followed by inversion of ② through the point shown on the

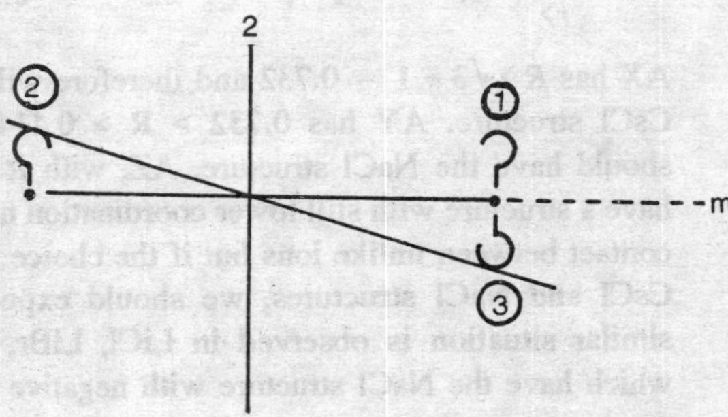

Fig. 9.9

2 axis gives ③. ③ is mirror image of ② reflected in the plane m perpendicular to the plane of the paper.

A-21. *(a)* See figure. *ABCD* is a suitable unit cell.

(b) Let r be the radius of a circle. The unit cell shown in the figure is composed of two equilateral triangles, *ABC*

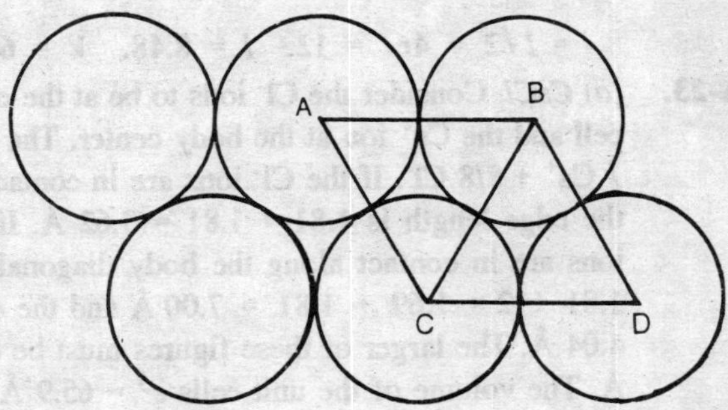

and *BCD*, each with side $2r$ and $s = 3r$; the area of the unit cell is $2(3r \cdot r^3)^{1/2} = 2r^2\sqrt{3}$. The unit cell contains $\frac{1}{6} + \frac{1}{3} + \frac{1}{6} + \frac{1}{3} = 1$ circle of area πr^2. The fractional void area is

$$\frac{2r^2\sqrt{3} - \pi r^2}{2r^2\sqrt{3}} = 0.093$$

A-22. *(a)*

	AX	AY	AZ
$\frac{r<}{r>} = R$	1	0.5	0.33

AX has $R > \sqrt{3} - 1 = 0.732$ and therefore will probably have CsCl structure. AY has $0.732 > R > 0.414 = \sqrt{2} - 1$, and should have the NaCl structure. AZ, with $R < 0.414$, might have a structure with still lower coordination number to permit contact between unlike ions but if the choice is only between CsCl and NaCl structures, we should expect the latter. A similar situation is observed in LiCl, LiBr, and LiI, all of which have the NaCl structure with negative ions in contact.

(b) Let l be the edge of the unit cell.

AX. Body diagonal $= l\sqrt{3} = 2r_+ + 2r_- = 4.$ $l = 2.31.$

$$V = l^3 = 12.3$$

AY. $l = 2r_+ + 2r_- = 2 + 4 = 6.$ $V = 216$

AZ. Here negative ions are in contact along the face diagonal. Face diagonal

$$= l\sqrt{2} = 4r_- = 12. \quad l = 8.48. \quad V = 609.8.$$

A-23. *(a) CsCl.* Consider the Cl^- ions to be at the corners of the unit cell and the Cs^+ ion at the body center. The unit cell contains *1* Cs^+ + 8/8 Cl^-. If the Cl^- ions are in contact along the edge, the edge length is 1.81 + 1.81 = 3.62 Å. If the Cs^+ and Cl^- ions are in contact along the body diagonal, this diagonal is $1.81 + 2 \times 1.69 + 1.81 = 7.00$ Å and the edge is $7.00/\sqrt{3} = 4.04$ Å. The larger of these figures must be correct : $e = 4.04$ Å. The volume of the unit cells $e^3 = 65.9$ Å^3. The volume of the ions is $\frac{4}{3}\pi(1.81^3 + 1.69^3) = 45.1$Å^3. The fractional void volume is $(65.9 - 45.1)/65.9 = 0.316$.

(b) NaCl. Unit cell contains $4Na^+ + 4Cl^-$. If Na^+ and Cl^- are in contact along an edge, the edge is $2(0.95 + 1.81) = 5.52$ Å. If Cl^- ions are in contact along a face diagonal, this diagonal is $4 \times 1.81 = 7.24$ Å and the edge is $7.24/\sqrt{2} = 5.12$ Å. If Na^+ and Cl^- are in contact along the body diagonal, this diagonal is $2(0.95 + 1.81) = 5.52$ Å, and the edge is $5.52/\sqrt{3} = 3.19$ Å.

The largest of these figure must be the correct one : e = 5.52 Å, e^3 = 168 Å^3. The volume of the ions is $4 \times \frac{4}{3}\pi\,(0.95^3 + 1.81^3) = 114$ Å^3.$(168 - 114)/168 = 0.32$.

(c) LiCl. Edge contact:$e = 4 \times 1.81/\sqrt{2} = 5.12$ Å. Body diagonal contact:$e = 4.82/\sqrt{3} = 2.78$ Å. Correct value : e = 5.12 Å. e^3 = 134 Å^3. Volume of ions $= 4 \times \frac{4}{3}\pi\,(0.68^3 + 1.81^3) = 103$ Å. $(134 - 103)/134 = 0.23$.

10

Short Answer Questions

Q. 10.1. Suggest explanations for the following statements:

(i) Sodium chloride pieces are harder than sodium metal;
(ii) Copper is ductile and malleable but brass is tenacious;
(iii) Latent heat of fusion of solid carbon dioxide is much less than that of silicon dioxide.

Ans. (*i*) Sodium chloride pieces are ionic crystals whereas sodium metal is a metallic crystal. In ionic crystals, the constituent ions are held together by strong electrostatic forces of attraction and thus the positions occupied by the ions are fixed. Hence the ionic crystal is hard.

In metallic crystal of sodium metal, the positive Na^+ ions are surrounded by a sea of electrons. Therefore, the positions of positive ions are not rigid. Hence the sodium metal is not hard enough.

(*ii*) Copper metal constitutes the metallic crystals in which the positive metal ions are traversed by electron gas. In copper metal, the positive metal ions are arranged in planes which impart malleability and ductility to the copper metal.

When copper is converted into brass, an alloy of copper and zinc, addition of Zn deforms simple planar structure of copper. Thus, alloy becomes tenacious.

(*iii*) Silicon dioxide is an example of covalent crystal in which atoms are linked together by a continuous chain of covalent bonds in fixed directions, thus giving it a giant structure. This is very hard and therefore its latent heat of fusion will be very high. On the other hand, solid carbon dioxide is a molecular crystal in which molecules are held together by weak van der Waal's forces and hence its latent heat of fusion is much less than that of silicon dioxide.

Q. 10.2. What is the importance of tetrahedral and octahedral holes in close packed stacks of spheres?

Ans. If atoms of other elements but having comparable size are available, they may fill interstices in crystals. For example, the atoms like B, C, H, N may fill interstices in lattice structures of metals and hence the resulting products will be very hard.

Q. 10.3. Which type of crystal should be the hardest and have the highest melting point?

Ans. Covalent crystals. The hardest substance known is diamond, a covalent crystal.

Q. 10.4. Diamond, by virtue of its structure, is a very hard substance. Suggest another substance which might be about as hard as diamond.

Ans. A substance isoelectronic with diamond, is BN, which, like carbon exists in two forms, one like graphite and one which has the diamond structure and which turns out to be slightly harder than diamond.

Q. 10.5. What is the effect of temperature on the conductivity of metal and semi-conductors?

Ans. If the temperature increases, conductivity of metals generally decreases while that of semi-conductor increases.

Q. 10.6. Will the energy released in crystallisation be equal to greater than or less than the energy absorbed in melting? Explain why?

Ans. Equal to. When the particles crystallise, the potential energy is released as heat which compensates for the energy removed.

Q. 10.7. What type of crystal is ice?

Ans. Molecular.

Q. 10.8. In ionic crystals, is the structure usually determined by the packing of cations around and anion or by anions around a cation.

Ans. Anions around a cation, since the cation will usually be the smaller.

Fill in the blanks :

Q. 1. In the following statements fill in the blanks choosing appropriate words from the list:

(*a*) crystal habits (*b*) amorphous
(*c*) allotropes (*d*) liquid crystals.

(*i*) Monoclinic sulphur and rhombic sulphur are two .... of sulphur.

(ii) A solid which melts over a broad range of temperatures is .... in nature.

(*iii*) Crystals can be classified into seven basic ....

(*iv*) If a crystalline pure solid on heating forms a turbid looking liquid which on further heating forms a clear liquid, the first liquid form is known as ....

Q. 2. The .... is the smallest fraction of the crystal lattice which, when repeated over and over in three dimensions, will reproduce the crystal.

Q. 3. NaF and NaCl have same crystal structure so that they are said to be ....

Q. 4. When a solid begins to melt, we sometimes say it is beginnings to ....

Q. 5. The formula for calcium fluoride is CaF_2 so there should be .... (how many?) fluoride ions in the unit cell.

Q. 6. The answer to the previous question suggests that the fluoride ions alone must assume a crystal lattice called ....

Q. 7. Each chlorine atom is surrounded by .... sodium atoms and each sodium ion is surrounded by .... chlorine atoms.

Q. 8. The existence of a substance in more than one crystalline form is called ....

Ans. 1. (*i*) Allotropes (*ii*) Amorphous (*iii*) Crystal habits (*iv*) Liquid crystal. 2. Unit cell 3. Isomorphous 4. Fuse 5. eight (twice as many as calcium) 6. simple cubic 7. six, six 8. polymorphism.

QUESTIONS

Essay Type

Q. 1. Differentiate between crystalline and amorphous solids. Explain the terms : space lattice and unit cell.

Q. 2. Describe briefly the structure of graphite. Why is graphite a good conductor of electricity?

Q. 3. Explain the meaning of elements of symmetry of a crystal. Describe the various elements of symmetry in a cubic crystal.

Q. 4. Describe the close packing of solid spheres. Explain the meaning of hexagonal close packing and cubic close packing.

Q. 5. What is meant by metallic bonding? How is it explained? Discuss the various theories. How would you account for the following :

(*a*) Metallic lustre (*b*) High thermal conductivity (*c*) Ductility (*d*) Thermal conductivity (*e*) Tensile strength (*f*) Softness (g) malleability.

Q. 6. Classify the crystals on the basis of the nature of forces which bind their constituent particles together.

Q. 7. Define lattice energy. How is it determined? What is Madelung constant?

Q. 8. What is Born-Haber cycle? How is it used for calculating lattice energy of sodium chloride?

Q. 9. Write down the Born-Lande equation and define the various terms used therein.

Q. 10. Explain the structures of NaCl, $CaCl_2$, ZnS, CdI_2, CaC_2 and CaF_2. Why is NaCl having an octahedral structure whereas

CsCl a body-centred cubic structure?

Q. 11. What is meant by coordination number? What types of crystal structure are associated with different coordination numbers? How are the minimum values of radius ratio arrived at for various coordination numbers and what are their limits?

Q. 12. What is meant by an ideal crystal? What is meant by Schottky defect and Frenkel defect? What are the consequences of these defects?

Q. 13. What is meant by non-stoichiometric defects of crystals. Explain the meaning of metal excess defects. What are the consequences of metal excess defects?

Short-Answer Type :

1. Is pure germanium a semi-conductor?

2. What type of semiconductors are used as rectifiers?

3. What is the effect of temperature on semiconduction?

4. What is meant by *n*-type and *p*-type semiconduction?

5. What are the consequences of metal deficiency defects?

6. How would you calculate lattice energy of ionic solids?

7. Explain why sodium chloride grows from water as cubes but from 15% aqueous urea solution as octahedra.

8. Why are ionic crystals hard, brittle and have high melting points?

9. What are semiconducting properties of silicon and germanium due to?

10. Which of the following substances will conduct electricity in solid state and why?

(*i*) Iodine (*ii*) Graphite
(*iii*) Diamond (*iv*) Mercury.